Thoma

*Graduate Texts in Mathematics* 17

Managing Editors: P. R. Halmos
C. C. Moore

*M. Rosenblatt*

# Random Processes

Second Edition

Springer-Verlag New York · Heidelberg · Berlin

## Murray Rosenblatt

University of California, San Diego
Department of Mathematics
La Jolla, California 92037

*Managing Editors*

### P. R. Halmos

Indiana University
Department of Mathematics
Swain Hall East
Bloomington, Indiana 47401

### C. C. Moore

University of California
at Berkeley
Department of Mathematics
Berkeley, California 94720

AMS Subject Classification (1970)
60A05, 60E05, 60F05, 60G10, 60G15, 60G25, 60G45, 60G50, 60J05,
60J10, 60J60, 60J75, 62M10, 62M15, 28A65

*Library of Congress Cataloging in Publication Data*

Rosenblatt, Murray.
 Random processes.
 (Graduate texts in mathematics, 17)
 Bibliography: p.
 1. Stochastic processes.  I. Title.  II. Series.
QA274.R64  1974  519.2  74–10956
ISBN 0-387-90085-3

First edition: published 1962, by Oxford University Press, Inc.

ISBN 0-387-90085-3 Springer-Verlag New York · Heidelberg · Berlin
ISBN 3-540-90085-3 Springer-Verlag Berlin · Heidelberg · New York

To My Brother and My Parents

# ACKNOWLEDGEMENT

I am indebted to D. Rosenblatt who encouraged me to write an introductory book on random processes. He also motivated much of my interest in functions of Markov chains. My thanks are due to my colleagues W. Freiberger and G. Newell who read sections of the manuscript and made valuable suggestions. I would especially like to acknowledge the help of J. Hachigian and T. C. Sun, who looked at the manuscript in some detail and made helpful comments on it. Thanks are due to Ezoura Fonseca for patient and helpful typing. This book was written with the support of the Office of Naval Research.

1962

This edition by Springer Verlag of *Random Processes* differs from the original edition of Oxford University Press in the following respects. Corrections have been made where appropriate. Additional remarks have been made in the notes to relate topics in the text to the literature dated from 1962 on. A chapter on martingales has also been added. K. S. Lii, M. Sharpe and R. A. Wijsman made a number of helpful suggestions. Neola Crimmins typed the changes in the manuscript.

1973

# CONTENTS

# RANDOM PROCESSES

# NOTATION

$A \cup B$ : the set of points belonging to either of the sets $A$ and $B$, usually called the union of $A$ and $B$.

$\bigcup_i A_i$ : the set of points belonging to any of the sets $A_i$.

$AB$ or
$A \cap B$ : the set of points belonging to both of the sets $A$ and $B$, usually called the product or intersection of the sets $A$ and $B$.

$\bigcap_i A_i$ : the set of points belonging to all the sets $A_i$.

$A - B$ : the set of points in $A$ but not in $B$, usually called the difference of the sets $A$ and $B$.

$A \ominus B$ : the set of points in $A$ or $B$ but not both, usually called the symmetric difference of the sets $A$ and $B$.

$x \in A$ : $x$ an element of the set $A$.

$o$ : $f(x) = o(g(x))$ as $x \to r$ if $\lim_{x \to r} f(x)/g(x) = 0$

$O$ : $f(x) = O(g(x))$ as $x \to r$ if $|f(x)/g(x)| \le K < \infty$ as $x \to r$.

$\approx$ : $f \approx g$ $f$ is approximately the same as $g$.

$\cong$ : $f(x) \cong g(x)$ as $x \to r$ if $\lim_{x \to r} f(x)/g(x) = 1$.

$x \to y+$ : $x$ approaches $y$ from the right.

$x \bmod r$
with $r > 0$ : $x \bmod r = x - mr$ where $mr$ is the largest multiple of $r$ less than or equal to $x$.

$\delta_{\lambda,\mu}$
(Kronecker
delta) : $\delta_{\lambda,\mu}$ is equal to one if $\lambda = \mu$ and zero otherwise.

$\mathrm{Re}\ a$ : real part of the complex number $a$.

$\{\alpha | \ \ldots \ \}$ : the set of $\alpha$ satisfying the condition written in the place indicated by the three dots.
If $\alpha$ is understood this may simply be written as $\{ \ \cdots \ \}$.

All formulas are numbered starting with (1) at the beginning of each section of each chapter. If a formula is referred to in the same section in which it appears, it will be referred to by number alone. If the formula appears in the same chapter but not in the same section, it will be referred to by number and letter of the section in which it appears. A formula appearing in a different chapter will be referred to by chapter, letter of section, and number. Suppose we are reading in section b of Chapter III. A reference to formula (13) indicates that the formula is listed in the same chapter and section. Formula (a.13) is in section a of the same chapter. Formula (II.a.13) is in section a of Chapter II.

# INTRODUCTION

**1**

This text has as its object an introduction to elements of the theory of random processes. Strictly speaking, only a good background in the topics usually associated with a course in Advanced Calculus (see, for example, the text of Apostol [1]) and the elements of matrix algebra is required although additional background is always helpful. Nonetheless a strong effort has been made to keep the required background on the level specified above. This means that a course based on this book would be appropriate for a beginning graduate student or an advanced undergraduate.

Previous knowledge of probability theory is not required since the discussion starts with the basic notions of probability theory. Chapters II and III are concerned with discrete probability spaces and elements of the theory of Markov chains respectively. These two chapters thus deal with probability theory for finite or countable models. The object is to present some of the basic ideas and problems of the theory in a discrete context where difficulties of heavy technique and detailed measure theoretic discussions do not obscure the ideas and problems. Further, the hope is that the discussion in the discrete context will motivate the treatment in the case of continuous state spaces on intuitive grounds. Of course, measure theory arises quite naturally in probability theory, especially so in areas like that of ergodic theory. However, it is rather extreme and in terms of motivation rather meaningless to claim that probability theory is just measure theory. The basic measure theoretic tools required for discussion in continuous state spaces are introduced in Chapter IV without proof and motivated on intuitive grounds and by comparison with the discrete case. For otherwise, we would get lost in the detailed derivations of measure theory. In fact, throughout the book the presentation is made with the main object understanding of the material on intuitive grounds. If rigorous proofs are proper and meaningful with this view in mind they are presented. In a number of places where such rigorous discussions are too lengthy and do not give much immediate understanding, they may be deleted with heuristic discussions given in their place. However, this will be indicated in the derivations. Attention has been paid to the question of motivating the material in terms of the situations in which the probabilistic problems dealt with typically arise.

The principal topics dealt with in the following chapters are strongly and weakly stationary processes and Markov processes. The basic result in the chapter on strongly stationary processes is the ergodic theorem. The related concepts of ergodicity and mixing are also considered. Fourier analytic methods are the appropriate tools for weakly stationary processes. Random harmonic analysis of these processes is considered at some length in Chapter VII. Associated statistical questions relating to spectral estimation for Gaussian stationary processes are also discussed. Chapter VI deals with Markov processes. The two extremes of jump processes and diffusion processes are dealt with. The discussion of diffusion processes is heuristic since it was felt that the detailed sets of estimates involved in a completely rigorous development were rather tedious and would not reward the reader with a degree of understanding consonant with the time required for such a development.

The topics in the theory of random processes dealt with in the book are certainly not fully representative of the field as it exists today. However, it was felt that they are representative of certain broad areas in terms of content and development. Further, they appeared to be most appropriate for an introduction. For extended discussion of the various areas in the field, the reader is referred to Doob's treatise [12] and the excellent monographs on specific types of processes and their applications.

As remarked before, the object of the book is to introduce the reader as soon as possible to elements of the theory of random processes. This means that many of the beautiful and detailed results of what might be called classical probability theory, that is, the study of independent random variables, are dealt with only insofar as they lead to and motivate study of dependent phenomena. It is hoped that the choice of models of random phenomena studied will be especially attractive to a student who is interested in using them in applied work. One hopes that the book will therefore be appropriate as a text for courses in mathematics, applied mathematics, and mathematical statistics. Various compromises have been made in writing the book with this in mind. They are not likely to please everyone. The author can only offer his apologies to those who are disconcerted by some of these compromises.

Problems are provided for the student. Many of the problems may be nontrivial. They have been chosen so as to lead the student to a greater understanding of the subject and enable him to realize the potential of the ideas developed in the text. There are references to the work of some of the people that developed the theory discussed. The references are by no means complete. However, I hope they do give some sense of historical development of the ideas and techniques as they exist today. Too often, one gets the impression that a body of theory has arisen instantaneously since the usual reference is given to the latest or most current version of that theory. References are also given to more extended developments of theory and its application.

Some of the topics chosen are reflections of the author's interest. This is perhaps especially true of some of the discussion on functions of Markov chains and the uniform mixing condition in Chapters III and IX. The section on functions of Markov chains does give much more insight into the nature of the Markov assumption. The uniform mixing condition is a natural condition to introduce if one is to have asymptotic normality of averages of dependent processes.

Chapter VIII has been added because of the general interest in martingales. Optional sampling and a version of a martingale convergence theorem are discussed. A central limit theorem for martingales is derived and applied to get a central limit theorem for stationary processes.

# II

# BASIC NOTIONS FOR FINITE
# AND DENUMERABLE STATE MODELS

## a. Events and Probabilities of Events

Let us first discuss the intuitive background of a context in which the probability notion arises before trying to formally set up a probability model. Consider an experiment to be performed. Some event $A$ may or may not occur as a result of the experiment and we are interested in a number $P(A)$ associated with the event $A$ that is to be called the probability of $A$ occurring in the experiment. Let us assume that this experiment can be performed again and again under the same conditions, each repetition independent of the others. Let $N$ be the total number of experiments performed and $N_A$ be the number of times event $A$ occurred in these $N$ performances. If $N$ is large, we would expect the probability $P(A)$ to be close to $N_A/N$

$$P(A) \approx N_A/N. \tag{1}$$

In fact, if the experiment could be performed again and again under these conditions without end, $P(A)$ would be thought of ideally as the limit of $N_A/N$, as $N$ increases without bound. Of course, all this is an intuitive discussion but it sets the framework for some of the basic properties one expects the probability of an event in an experimental context to have. Thus $P(A)$, the probability of the event $A$, ought to be a real number greater than or equal to zero and less than or equal to 1

$$0 \leq P(A) \leq 1. \tag{2}$$

Now consider an experiment in which two events $A_1$, $A_2$ might occur. Suppose we wish to consider the event "either $A_1$ or $A_2$ occurs," which we shall denote notationally by $A_1 \cup A_2$. Suppose the two events are disjoint in the following sense: the event $A_1$ can occur and the event $A_2$ can occur but both cannot occur simultaneously. Now consider repeating the same experiment independently a large number of times, say $N$.

Then intuitively

$$P(A_1) \approx N_{A_1}/N, \qquad P(A_2) \approx N_{A_2}/N,$$
$$P(A_1 \cup A_2) \approx N_{A_1 \cup A_2}/N. \qquad (3)$$

But $N_{A_1 \cup A_2}$, the number of times "$A_1$ or $A_2$ occurs" in the experiment is equal to $N_{A_1} + N_{A_2}$. Thus if $A_1$, $A_2$ are disjoint we ought to have

$$P(A_1 \cup A_2) = P(A_1) + P(A_2). \qquad (4)$$

By extension, if a finite number of events $A_1, \ldots, A_n$ can occur in an experiment, let $A_1 \cup A_2 \cup \cdots \cup A_n = \bigcup_{i=1}^{n} A_i$ denote the event "either $A_1$ or $A_2$ or $\ldots$ or $A_n$ occurs in the experiment." If the events are disjoint, that is, no two can occur simultaneously, we anticipate as before that

$$P(\bigcup_{i=1}^{n} A_i) = \sum_{i=1}^{n} P(A_i). \qquad (5)$$

Of course, if the events are not disjoint such an additivity relation will not hold. The notation $\cup A_i$ need not be restricted to a finite collection of events $\{A_i\}$. It will also be used for infinite collections of events. Relation (5) would be expected to hold for a denumerable or countable collection $A_1, A_2, \ldots$ of disjoint events.

There is an interesting but trivial event $\Omega$, the event "something occurs." It is clear that $N_\Omega = N$ and hence

$$P(\Omega) = 1. \qquad (6)$$

With each event $A$ there is associated an event $\bar{A}$, "$A$ does not occur." We shall refer to this event as the complement of $A$. Since $N_{\bar{A}} = N - N_A$ it is natural to set

$$P(\bar{A}) = 1 - P(A). \qquad (7)$$

Notice that the complement of $\Omega$, $\phi = \bar{\Omega}$ ("nothing occurs") has probability zero

$$P(\phi) = 1 - P(\Omega) = 0. \qquad (8)$$

Let us now consider what is implicit in our discussion above. A family of events is associated with the experiment. The events represent classes of outcomes of the experiment. Call the family of events $A$ associated with the experiment $\mathfrak{F}$. The family of events $\mathfrak{F}$ has the following properties:

*1 1. If the events $A_1$, $A_2 \epsilon \mathfrak{F}$ then the event $A_1 \cup A_2$, "either $A_1$ or $A_2$ occurs," is an element of $\mathfrak{F}$.*

2. *The event $\Omega$, "something occurs" is an element of $\mathfrak{F}$.*

3. *Given any event $A\epsilon\mathfrak{F}$, the complementary event $\bar{A}$, "$A$ does not occur,"
is an element of $\mathfrak{F}$.*

Further, a function of the events $A\epsilon\mathfrak{F}$, $P(A)$, is given with the following
properties:

2  1. $0 \leq P(A) \leq 1$

   2. $P(\Omega) = 1$

   3. $P(A_1 \cup A_2) = P(A_1) + P(A_2)$ *if* $A_1$, $A_2\epsilon\mathfrak{F}$ *are disjoint.*

Notice that the relation

$$P(\bar{A}) = 1 - P(A) \tag{9}$$

follows from 2.2 and 2.3.

In the case of an experiment with a finite number of possible elementary outcomes we can distinguish between compound and simple
events associated with the experiment. A simple event is just the specification of a particular elementary outcome. A compound event is the
specification that one of several elementary outcomes has been realized
in the experiment. Of course, the simple events are disjoint and can be
thought of as sets, each consists of one point, the particular elementary
outcome each corresponds to. The compound events are then sets each
consisting of several points, the distinct elementary outcomes they
encompass. In the probability literature the simple events are at times
referred to as the "sample points" of the probability model at hand.
The probabilities of the simple events, let us say $E_1$, $E_2$, $\ldots$ , $E_n$, are
assumed to be specified. Clearly

$$0 \leq P(E_i) \leq 1 \tag{10}$$

and since the simple events are disjoint and exhaustive (in that they
account for all possible elementary outcomes of the experiment)

$$\sum_{i=1}^{n} P(E_i) = 1. \tag{11}$$

The probability of any event $A$ by 2.3 is

$$P(A) = \sum_{E_i \subset A} P(E_i). \tag{12}$$

The events $A$ of $\mathfrak{F}$ are the events obtained by considering all possible
collections of elementary occurrences. Thus the number of distinct
events $A$ of $\mathfrak{F}$ are $2^n$ altogether. A collection of events (or sets) satisfying
conditions *1.1–1.3* is commonly called a *field*. In the case of experiments

with an infinite number of possible elementary outcomes one usually wishes to strengthen assumption *1* in the following way:

*1 1'. Given any denumerable* (finite or infinite) *collection of events $A_1$, $A_2$, . . . of $\mathfrak{F}$ $A_1 \cup A_2 \cup \cdot \cdot \cdot = \cup A_i$ "either $A_1$ or $A_2$ or . . . occurs" is an element of $\mathfrak{F}$.* Such a collection of events or sets with property *1.1* replaced by *1.1'* is called a *sigma-field*. In dealing with $P$ as a function of events $A$ of a *$\sigma$-field $\mathfrak{F}$*, assumption *2.3* is strengthened and replaced by

$$2.3' \quad P(\cup A_i) = \sum_i P(A_i) \text{ if } A_1, A_2, \ldots, \epsilon \mathfrak{F} \tag{13}$$

*is a denumerable collection of disjoint events.* This property is commonly referred to as countable additivity of the $P$ function.

By introducing "sample points" we are able to speak alternatively of events or sets. In fact disjointness of events means disjointness of the corresponding events viewed as collections of elementary outcomes of the experiment. Generally, it will be quite convenient to think of events as sets and use all the results on set operations which have complete counterparts in operations on events. In fact the $\cup$ operation on events is simply set addition for the events regarded as sets. Similarly complementation of an event amounts to set complementation for the event regarded as a set.

It is very important to note that our basic notion is that of an experiment with outcomes subject to random fluctuation. A family or field of events representing the possible outcomes of the experiment is considered with a numerical value attached to each event. This numerical value or probability associated with the event represents the relative frequency with which one expects the event to occur in a large number of independent repetitions of the experiment. This mode of thought is very much due to von Mises [57].

Let us now illustrate the basic notions introduced in terms of a simple experiment. The experiment considered is the toss of a die. There are six elementary outcomes of the experiment corresponding to the six faces of the die that may face up after a toss. Let $E_i$ represent the elementary event "$i$ faces up on the die after the toss." Let

$$0 \leq p_i = P(E_i) \leq 1 \tag{14}$$

be the probability of $E_i$. The probability of the compound event $A = \{$an even number faces up$\}$ is easily seen to be

$$P(A) = p_2 + p_4 + p_6. \tag{15}$$

The die is said to be a "fair" die if

$$p_1 = p_2 = \cdots = p_6 = \tfrac{1}{6}.$$

Another event or set operation that is of importance can be simply derived from those already considered. Given two events $A_1$, $A_2 \epsilon \mathfrak{F}$, consider the derived event $A_1 \cap A_2$ "both $A_1$ and $A_2$ occur." It is clear that

$$A_1 A_2 = A_1 \cap A_2 = (\overline{\bar{A}_1 \cup \bar{A}_2}). \tag{16}$$

# b.  Conditional Probability, Independence, and Random Variables

A natural and important question is what is to be meant by the conditional probability of an event $A_1$ given that another event $A_2$ has occurred. The events $A_1$, $A_2$ are, of course, possible outcomes of a given experiment. Let us again think in terms of a large number $N$ of independent repetitions of the experiment. Let $N_{A_2}$ be the number of times $A_2$ has occurred and $N_{A_1 \cap A_2}$ the number of times $A_1$ and $A_2$ have simultaneously occurred in the $N$ repetitions of the experiment. It is quite natural to think of the conditional probability of $A_1$ given $A_2$, $P(A_1|A_2)$, as very close to

$$N_{A_1 \cap A_2}/N_{A_2} = \frac{N_{A_1 \cap A_2}}{N} \Big/ \frac{N_{A_2}}{N} \tag{1}$$

if $N$ is large. This motivates the definition of the conditional probability $P(A_1|A_2)$ by

$$P(A_1|A_2) = P(A_1 \cap A_2)/P(A_2) \tag{2}$$

which is well defined as long as $P(A_2) > 0$. If $P(A_2) = 0$, $P(A_1|A_2)$ can be taken as any number between zero and one. Notice that with this definition of conditional probability, given any $B \epsilon \mathfrak{F}$ (the field of events of the experiment) for which $P(B) > 0$, the conditional probability $P(A|B)$, $A \epsilon \mathfrak{F}$, as a function of $A \epsilon \mathfrak{F}$ is a well-defined probability function satisfying 2.1-2.3. It is very easy to verify that

$$\sum_i P(A|E_i)P(E_i) = P(A) \tag{3}$$

where the $E_i$'s are the simple events of the probability field $\mathfrak{F}$. A similar relation will be used later on to define conditional probabilities in the case of experiments with more complicated spaces of sample points (sample spaces).

The term independence has been used repeatedly in an intuitive and unspecified sense. Let us now consider what we ought to mean by the independence of two events $A_1$, $A_2$. Suppose we know that $A_2$ has occurred. It is then clear that the relevant probability statement about $A_1$ is a statement in terms of the conditional probability of $A_1$ given $A_2$. It would be natural to say that $A_1$ is independent of $A_2$ if the conditional probability of $A_1$ given $A_2$ is equal to the probability of $A_1$

$$P(A_1|A_2) = P(A_1), \tag{4}$$

that is, the knowledge that $A_2$ has occurred does not change our expectation of the frequency with which $A_1$ should occur. Now

$$P(A_1|A_2) = P(A_1 \cap A_2)/P(A_2) = P(A_1)$$

so that

$$P(A_1 \cap A_2) = P(A_1)P(A_2). \tag{5}$$

Note that the argument phrased in terms of $P(A_2|A_1)$ would lead to the same conclusion, namely relation (5). Suppose a denumerable collection (finite or infinite) of events $A_1$, $A_2$ . . . is considered. We shall say that the collection of events is a collection of independent events if every finite subcollection of events $A_{k_1}$, . . . , $A_{k_m}$, $1 \leq k_1 < \cdots < k_m$, satisfies the product relation

$$P(A_{k_1}A_{k_2} \cdots A_{k_m}) = \prod_{i=1}^{m} P(A_{k_i}).$$

It is easy to give an example of a collection of events that are pairwise independent but not jointly independent. Let $\mathfrak{F}$ be a field of sets with four distinct simple events $E_1$, $E_2$, $E_3$, $E_4$

$$P(E_i) = \tfrac{1}{4}, \quad i = 1, \ldots, 4. \tag{6}$$

Let the compound events $A_i$  $i = 1, 2, 3$ be given by

$$A_1 = E_1 \cup E_2$$
$$A_2 = E_1 \cup E_3$$
$$A_3 = E_1 \cup E_4.$$

Then

$$P(A_i) = \tfrac{1}{2} \quad i = 1, 2, 3 \tag{7}$$

while

$$P(A_1A_2) = P(A_1A_3) = P(A_2A_3) = P(E_1) = \tfrac{1}{4}. \tag{8}$$

The events $A_i$ are clearly pairwise independent. Nonetheless

$$P(A_1A_2A_3) = P(E_1) = \tfrac{1}{4} \neq P(A_1)P(A_2)P(A_3). \tag{9}$$

Thus far independence of events within a collection has been discussed. Suppose we have several collections of events $C_1 = \{A_i^{(1)}; i = 1, \ldots, n_1\}$, $C_2 = \{A_i^{(2)}; i = 1, \ldots, n_2\}$, $\ldots$, $C_m = \{A_i^{(m)}; i = 1, \ldots, n_m\}$. What shall we mean by the independence of these collections of events? It is natural to call the collections $C_1, \ldots, C_m$ independent if every $m$-tuple of events $A_{i_1}^{(1)}, \ldots, A_{i_m}^{(m)}$ consisting of one event from each collection is a collection of independent events.

This discussion of independence of collections of events can now be applied in defining what we ought to mean by independence of experiments. Suppose we have $m$ experiments with corresponding fields $\mathfrak{F}_1, \ldots, \mathfrak{F}_m$. Let the corresponding collections of simple events be

$$\{E_i^{(1)}; i = 1, \ldots, n_1\}, \ldots, \{E_i^{(m)}; i = 1, \ldots, n_m\}.$$

Now the $m$ experiments can be considered jointly as one global experiment in which case the global experiment has a field of events generated by the following collection of simple events

$$E_{i_1, \ldots, i_m} = E_{i_1}^{(1)} E_{i_2}^{(2)} \cdots E_{i_m}^{(m)} \tag{10}$$

and the $m$ experiments are said to be independent if

$$P(E_{i_1, \ldots, i_m}) = P(E_{i_1}^{(1)} E_{i_2}^{(2)} \cdots E_{i_m}^{(m)}) = \prod_{k=1}^{m} P(E_{i_k}^{(k)}). \tag{11}$$

Consider this in the case of a simple coin tossing experiment. The coin has two faces, head and tail, denoted by 1 and 0 respectively. The probability of a head in a coin toss is $p$, $0 \leq p \leq 1$. Suppose the coin is tossed $m$ times, each time independent the others. Each coin toss can be regarded as an experiment, in which case we have $m$ independent experiments. If the $m$ experiments are jointly regarded as one experiment, each simple event can be represented as

$$E_{i_1, \ldots, i_m} = \{(i_1, \ldots, i_m)\}, \qquad i_1, \ldots, i_m = 0, 1. \tag{12}$$

Thus each simple event consists of one point, an $m$-vector with coordinates 0 or 1. Each such point is a sample point. Since the coin tosses are independent

$$P(E_{i_1, \ldots, i_m}) = P\{(i_1, \ldots, i_m)\} = \prod_{k=1}^{m} P(E_{i_k}^{(k)}) = p^{\Sigma i_k} q^{m - \Sigma i_k} \tag{13}$$

where $q = 1 - p$. If the coin is fair, that is, $p = q = \frac{1}{2}$, the probabilities of simple events are all equal to $\frac{1}{2}^m$.

We can regard the models of experiments dealt with as triplets of entities $(\Omega, \mathfrak{F}, P)$ where $\Omega$ is a space of points (all the sample points),

$\mathfrak{F}$ the field (if there are a finite number of sample points) or sigma-field (if there are a denumerably infinite number of sample points) of events generated by the sample points, and $P$ is the probability function defined on the events of $\mathfrak{F}$. Such a model of an experiment is called *a probability space*. Usually the sample points are written as $w$. A numerical valued function $X(w)$ on the space $\Omega$ of sample points is called a random variable. Thus $X(w)$ represents an observable in the experiment. In the case of the $m$ successive independent coin tossings discussed above, the number of heads obtained would be a random variable. A random variable $X(w)$ generates a field (sigma-field) $\mathfrak{F}_X$ of events generated by events of the form $\{w|X(w) = a\}$ where $a$ is any number. The field consists of events which are unions of events of the form $\{w|X(w) = a\}$. The probability function $P$ on the events of this field $\mathfrak{F}_X$ generated by $X(w)$ is called the *probability distribution of $X(w)$*. Quite often the explicit indication of $X(w)$ as a function of $w$ is omitted and the random variable $X(w)$ is written as $X$. We shall typically follow this convention unless there is an explicit need for clarification. Suppose we have $n$ random variables $X_1(w), \ldots , X_n(w)$ defined on a probability space. The random variables $X_1, \ldots , X_n$ *are said to be independent* if the fields (sigma-fields) $\mathfrak{F}_{X_1}, \ldots , \mathfrak{F}_{X_n}$ generated by them are independent.

The discussion of a probability space and of random variables on the space is essentially the same in the case of a sample space with a nondenumerable number of sample points. The discussion must, however, be carried out much more carefully due to the greater complexity of the context at hand. We leave such a discussion for Chapter IV.

## c. The Binomial and Poisson Distributions

Two classical probability distributions are discussed in this section. The first distribution, the binomial, is simply derived in the context of the coin tossing experiment discussed in the previous section. Consider the random variable $X = \{$number of heads in $m$ successive independent coin tossings$\}$. Each sample point $(i_1, \ldots , i_m)$, $i_k = 0, 1$, of the probability space corresponding to an outcome with $r$ heads and $m - r$ tails, $0 \leq r \leq m$, has probability $p^r q^{m-r}$ where $q = 1 - p$, $0 \leq p \leq 1$. But there are precisely factorial coefficient

$$\binom{m}{r} = \frac{m!}{r!(m - r)!} \tag{1}$$

such distinct sample points with $r$ heads and $m - r$ tails. Therefore the probability distribution of $X$ is given by

$$P(X = r) = \binom{m}{r} p^r q^{m-r} \qquad r = 0, 1, \ldots , m. \tag{2}$$

Of course,

$$\sum_{r=0}^{m} P(X = r) = \sum_{r=0}^{m} \binom{m}{r} p^r q^{m-r} = 1 \tag{3}$$

and we recognize the probabilities as the terms in the binomial expansion

$$(p + q)^m = \sum_{r=0}^{m} \binom{m}{r} p^r q^{m-r}, \tag{4}$$

an obvious motivation for the name binomial distribution.

The Poisson distribution is obtained from the binomial distribution by a limiting argument. Set $mp = \lambda > 0$ with $\lambda$ constant and consider

$$\lim_{m \to \infty} P(X = r). \tag{5}$$

Now

$$\begin{aligned}
P(X = r) &= \binom{m}{r} p^r q^{m-r} \\
&= \frac{m(m-1) \cdots (m-r+1)}{r!} \left(\frac{\lambda}{m}\right)^r \left(1 - \frac{\lambda}{m}\right)^{m-r} \\
&= \frac{\lambda^r}{r!} \left(1 - \frac{\lambda}{m}\right)^m \left(1 - \frac{1}{m}\right) \cdots \left(1 - \frac{r-1}{m}\right) \left(1 - \frac{\lambda}{m}\right)^{-r} \\
&\to \frac{\lambda^r}{r!} e^{-\lambda}
\end{aligned} \tag{6}$$

as $m \to \infty$. A random variable $Y$ with probability distribution

$$P(Y = r) = \frac{\lambda^r}{r!} e^{-\lambda} \tag{7}$$

is said to have a Poisson distribution. It is clear that we would expect this distribution to be a good approximation when the experiment can be regarded as a succession of many independent simple binomial trials (a simple binomial trial is an experiment with a simple success or failure outcome), the probability of success $p = \dfrac{\lambda}{m}$ is small, and the probability distribution of the total number of successes is desired.

Such is the case when dealing with a Geiger counter for radioactive material. For if we divide the time period of observation into many small equal subintervals, the over-all experiment can then be regarded as an ensemble of independent binomial experiments, one corresponding to each subinterval. In each subinterval there is a large probability $1 - \frac{\lambda}{m}$ that there will be no scintillation and a small probability $\frac{\lambda}{m}$ that there will be precisely one scintillation.

# d. Expectation and Variance of Random Variables (Moments)

Let $X$ be a random variable on a probability space with probability distribution

$$P(X = a_i) = p_i \qquad i = 1, 2, \ldots \tag{1}$$

The *expectation* of $X$, that is, $EX$, will be defined for random variables $X$ on the probability space with

$$\sum_{i=1}^{\infty} |a_i| p_i \tag{2}$$

finite. As we shall see, $E$ can be regarded as a linear operator acting on these random variables. The expectation $EX$ is defined as

$$EX = \sum_{i=1}^{\infty} a_i p_i. \tag{3}$$

Thus $EX$ is just the mean or first moment of the probability distribution of $X$. More generally, $n$-th order moments, $n = 0, 1, \ldots$, are defined for random variables $X$ with

$$\sum_{i=1}^{\infty} |a_i|^n p_i < \infty. \tag{4}$$

The $n$-th order moment of $X$ is defined as the expectation of $X^n$, $EX^n$,

$$EX^n = \sum_{i=1}^{\infty} a_i^n p_i. \tag{5}$$

The $n$-th order absolute moment of $X$ is

$$E|X|^n = \sum_{i=1}^{\infty} |a_i|^n p_i. \tag{6}$$

The first moment or mean of $X$, $m = EX$, is the center of mass of the probability distribution of $X$, where probability is regarded as mass. Let $X$, $Y$ be two random variables with well-defined expectations, $EX$, $EY$, and $\alpha$, $\beta$ any two numbers. Let the values assumed by $X$, $Y$ with positive probability be $a_i$, $b_i$ respectively. Then

$$E(\alpha X + \beta Y) = \sum_{i,j} (\alpha a_i + \beta b_j) P(X = a_i, Y = b_j)$$
$$= \alpha \sum_i a_i P(X = a_i) + \beta \Sigma b_i P(Y = b_i) \qquad (7)$$
$$= \alpha EX + \beta EY.$$

Thus $E$ is a linear operator on the random variables $X$ for which $EX$ is well defined. Of course, this can be extended to any finite number of such random variables $X_1, \ldots, X_m$ so that we have

$$E(\sum_{i=1}^{m} \alpha_i X_i) = \sum_{i=1}^{m} \alpha_i EX_i. \qquad (8)$$

It is easy to give an example of a random variable for which the expectation is undefined. Simply take $a_i = i, i = 1, 2, \ldots$ and set

$$p_i = Ki^{-3/2} \quad i = 1, 2, \ldots$$
$$K = (\sum_{i=1}^{\infty} i^{-3/2})^{-1}. \qquad (9)$$

Since

$$\sum_{i=1}^{\infty} ip_i = \sum_{i=1}^{\infty} i^{-1/2}K = \infty \qquad (10)$$

$EX$ is not well defined. This is due to the fact that too much probability mass has been put in the tail (large values of $X$) of the probability distribution of $X$.

Now consider two *independent* random variables $X,Y$ whose expectations are well defined. As before let the values assumed by $X,Y$ with positive probability be $a_i$, $b_i$ respectively. Then the expectation of the product $XY$ is given by

$$EXY = \sum_{i,j} a_i b_j P(X = a_i, Y = b_j)$$
$$= \sum_{i,j} a_i b_j P(X = a_i) P(Y = b_j) \qquad (11)$$
$$= E(X)E(Y).$$

Thus the expectation operator is multiplicative when dealing with products of independent random variables. If $X,Y$ are independent and $f,g$ are any two functions, $f(X),g(Y)$ are independent. The argument given above then indicates that

$$E(f(X)g(Y)) = Ef(X)Eg(Y) \tag{12}$$

if $Ef(X),Eg(Y)$ are well defined. This basic and important property will be used often when dealing with independent random variables.

A measure of concentration of the probability mass of a random variable $X$ about its mean is given by the central moment

$$\sigma^2 = E(X - m)^2 = E(X^2 - 2mX + m^2) \\ = EX^2 - m^2, \tag{13}$$

commonly called the variance of the probability distribution. The variance $\sigma^2(X) = \sigma^2$ is well defined as long as $EX^2$ is. The central moments are moments about the mean of the probability distribution. Just as in the case of noncentral moments, one can consider central moments (if any exist) of all non-negative integral orders

$$E(X - m)^n \quad n = 0, 1, 2, \ldots . \tag{14}$$

It is clear that

$$E(X - m)^0 = E1 = 1 \\ E(X - m) = 0 \\ E(X - m)^2 = \sigma^2 \tag{15}$$

$$\cdots .$$

There is a very interesting additive property of the variance in the case of independent random variables. Let $X_1, \ldots, X_s$ be independent random variables with finite second moments. Set

$$m_i = EX_i, \quad \sigma_i^2 = \sigma^2(X_i), \quad i = 1, \ldots, s. \tag{16}$$

Then the variance of the sum

$$\sigma^2 \left( \sum_1^s X_i \right) = E \left( \sum_1^s (X_i - m_i) \right)^2$$

$$= \sum_{i,j=1}^s E[(X_i - m_i)(X_j - m_j)]$$

$$= \sum_{i=1}^s \sigma_i^2 + \sum_{i \neq j} E(X_i - m_i)(X_j - m_j) \tag{17}$$

$$= \sum_{i=1}^s \sigma_i^2$$

by the independence of the random variables.

Let us now consider computing the first few moments of the binomial and Poisson distributions. First of all, by making use of (6) it is seen that all moments of these distributions are well defined. The moments will be evaluated by making use of a tool that is very valuable when dealing with probability distributions concentrated on the non-negative integers. A transform of the probability distribution commonly called the *generating function* of the distribution is introduced as follows

$$g(s) = \sum_{k=0}^{\infty} p_k s^k = E(s^X). \tag{18}$$

The generating function $g(s)$ is the formal power series with coefficient of $s^k$ the probability $p_k$. This power series is well defined on the closed interval $|s| \leq 1$ and infinitely differentiable on the open interval $|s| < 1$ since

$$p_k \geq 0, \ \Sigma p_k = 1. \tag{19}$$

Here all the moments $EX^n$ are absolute moments since the probability mass is concentrated on the non-negative integers. Certain moments, called factorial moments, are very closely related to the ordinary moments and can readily be derived from the generating function by differentiation. The $r$-th factorial moment of $X$

$$E[X(X-1) \cdots (X-r+1)] \quad r = 1, 2, \ldots \tag{20}$$

is well defined if and only if the $r$-th moment of $X$, $EX^r$, is well defined. Notice that

$$E[X(X-1) \cdots (X-r+1)] = \sum_{k=0}^{\infty} k(k-1) \cdots (k-r+1)p_k \tag{21}$$

$$= \lim_{s \to 1-} \sum_{k=0}^{\infty} k(k-1) \cdots (k-r+1)p_k s^{k-r} = \lim_{s \to 1-} \frac{d^r}{ds^r} g(s).$$

Here $s \to 1-$ indicates that $s$ approaches 1 from the left.

Let us now consider computing the moments of the binomial and Poisson distribution. First consider the binomial distribution. Its generating function

$$g(s) = \sum_{k=0}^{n} \binom{n}{k} p^k q^{n-k} s^k \tag{22}$$

$$= (ps + q)^n.$$

The $r$-th derivative

$$\frac{d^r}{ds^r} g(s) = n(n-1) \cdots (n-r+1)(ps+q)^{n-r}p^r \tag{23}$$

so that

$$\lim_{s \to 1-} \frac{d^r}{ds^r} g(s) = n(n-1) \cdots (n-r+1)p^r. \tag{24}$$

The first and second moments are given by

$$
\begin{aligned}
EX &= np \\
EX^2 &= E[X(X-1)] + EX \\
&= n(n-1)p^2 + np.
\end{aligned} \tag{25}
$$

The variance of the distribution

$$
\begin{aligned}
\sigma^2 &= EX^2 - (EX)^2 \\
&= n(n-1)p^2 + np - n^2p^2 \\
&= npq.
\end{aligned} \tag{26}
$$

The generating function of the Poisson distribution

$$g(s) = \sum_{k=0}^{\infty} \frac{\lambda^k}{k!} e^{-\lambda} s^k = e^{\lambda(s-1)}. \tag{27}$$

The $r$-th factorial moment

$$\lim_{s \to 1-} \frac{d^r}{ds^r} e^{\lambda(s-1)} = \lambda^r. \tag{28}$$

The first and second moments

$$
\begin{aligned}
EX &= \lambda \\
EX^2 &= \lambda^2 + \lambda
\end{aligned} \tag{29}
$$

so that the variance

$$\sigma^2 = \lambda \tag{30}$$

is equal to the mean.

Let $X$, $Y$ be two independent non-negative integer-valued random variables with probability distributions

$$P[X = k] = p_k, \qquad P[Y = k] = q_k \tag{31}$$

respectively, where $k = 0, 1, 2, \ldots$ . The generating function $h(s)$ of the sum $X + Y$ of the two random variables is readily given in terms of the generating functions $f(s)$, $g(s)$ of $X$ and $Y$ respectively. For

$$h(s) = E(s^{X+Y}) = E(s^X)E(s^Y) = f(s)g(s). \tag{32}$$

The probability distribution of $X + Y$ is given in terms of an operation on the $p$ and $q$ sequences commonly referred to as the convolution operation

$$P[X + Y = k] = \sum_{j=0}^{k} p_j q_{k-j} = (p*q)_k. \tag{33}$$

# e. The Weak Law of Large Numbers and the Central Limit Theorem

A very simple limit argument was used in section c to derive the Poisson distribution. This was a simple example of a limit theorem. In fact much of the classical literature in probability theory (which is primarily concerned with a study of independent random variables) is centered about such limit theorems. The weak law of large numbers and the central limit theorem are further examples of such limit theorems.

A simple but basic inequality due to Chebyshev is a necessary preliminary to our proof of the weak law of large numbers. *Let $X$ be a random variable with finite second moment. Then, given any positive number $\varepsilon(>0)$,*

$$P(|X| \geq \varepsilon) \leq EX^2/\varepsilon^2. \tag{1}$$

The proof is rather straightforward. For

$$
\begin{aligned}
EX^2 &= \sum_{i=1}^{\infty} a_i^2 p_i \\
&\geq \sum_{|a_i| \geq \varepsilon} a_i^2 p_i \\
&\geq \varepsilon^2 \sum_{|a_i| \geq \varepsilon} p_i = \varepsilon^2 P(|X| \geq \varepsilon).
\end{aligned}
\tag{2}
$$

This inequality gives us a crude but interesting estimate of the probability mass in the tail of the probability distribution in terms of the second moment of the distribution.

The weak law of large numbers follows. *Let $X_1, \ldots, X_n$ be independent random variables with the same probability distribution* (identically distributed) *and finite second moment.* Set

$$
\begin{aligned}
S_n &= \sum_{j=1}^{n} X_j \\
m &= EX_j \quad j = 1, \ldots, n.
\end{aligned}
\tag{3}
$$

*Then, given any* $\varepsilon > 0$,

$$P\left(\left|\frac{S_n}{n} - m\right| \geq \varepsilon\right) \to 0 \tag{4}$$

*as* $n \to \infty$. This states that for any small fixed positive number $\varepsilon$, there is an $n$ large enough so that most of the probability mass of the distribution of $S_n/n$ falls in the closed interval $|x - m| \leq \varepsilon$. The random variables $X_1, \ldots, X_n$ can be regarded as the observations in $n$ independent repetitions of the same experiment. In that case $S_n/n$ is simply the sample mean and the weak law of large numbers states that the mass of the probability distribution of the sample mean concentrates about the population mean $m = EX$ as $n \to \infty$. Intuitively, this motivates taking the sample mean as an estimate of the population mean when the sample size (number of experiments) is large. As we shall later see, it is essential that there be some moment condition such as that given in the statement of the weak law, that is, a condition on the amount of mass in the tail of the probability distribution of $X$. The law is called a weak law of large numbers because (4) amounts to a weak sort of convergence of $S_n/n$ to $m$. This point will be clarified later on in Chapter IV.

Now consider the proof of the law of large numbers. Let $\sigma^2$ be the common variance of the random variables $X_i$. Note that

$$P\left(\left|\frac{S_n}{n} - m\right| \geq \varepsilon\right) = P\left(\left|\sum_{i=1}^{n} (X_i - m)/n\right| \geq \varepsilon\right)$$

$$\leq E\left[\sum_{i=1}^{n} (X_i - m)/n\right]^2 \Big/ \varepsilon^2 = \frac{\sigma^2}{n\,\varepsilon^2} \tag{5}$$

by the Chebyshev inequality and the independence of the random variables $X_i$. On letting $n \to \infty$, we obtain the desired result.

We give a simple and exceedingly clever proof of the Weierstrass approximation theorem due to S. Bernstein [3]. This interpolation is appropriate because it indicates how probabilistic ideas at times lead to new approaches to nonprobabilistic problems. Consider the continuous functions on any *closed finite interval*. For convenience take the interval as [0,1]. The Weierstrass approximation theorem states that *any given continuous function on* [0,1] *can be approximated arbitrarily well uniformly on* [0,1] *by a polynomial of sufficiently high degree.* Serge Bernstein gave an explicit construction by means of his "Bernstein polynomials."

Let $f(x)$, $0 \leq x \leq 1$, be the given continuous function. Let $Y$ be a binomial variable of sample size $n$, that is, with $n$ coin tosses where the probability of success in one toss is $x$. Consider the derived random variable $f(Y/n)$. We might regard $f(Y/n)$ as an estimate of $f(x)$. This estimate is equal to $f(k/n)$, $k = 0, 1, \ldots, n$, with probability $\binom{n}{k} x^k (1 - x)^{n-k}$. As $n \to \infty$, by the weak law of large numbers, $Y/n$ approaches $x$ in probability and hence by the continuity of the function $f$, $f(Y/n)$ approaches $f(x)$ in probability. However, we are not really interested in $f(Y/n)$ but rather its mean value $Ef(Y/n)$. The expectation

$$Ef(Y/n) = \sum_{k=0}^{n} f(k/n) \binom{n}{k} x^k (1 - x)^{n-k} = p_n(x) \qquad (6)$$

is a polynomial of degree $n$ in $x$ which we shall call the Bernstein polynomial of degree $n$ corresponding to $f(x)$. A simple argument using the law of large numbers will show that $p_n(x)$ approaches $f(x)$ uniformly as $n \to \infty$. Since $f(x)$ is continuous on the closed interval $[0,1]$, it is uniformly continuous on $[0,1]$. Given any $\varepsilon > 0$, there is a $\delta(\varepsilon) > 0$ such that for any $x, y \epsilon [0,1]$ with $|x - y| < \delta(\varepsilon)$, $|f(x) - f(y)| < \varepsilon$. Consider any $\varepsilon > 0$. We shall show that for sufficiently large $n$

$$|p_n(x) - f(x)| < \varepsilon \qquad (7)$$

for all $x$. Note that

$$P\left[\left|\frac{Y}{n} - x\right| \geq \eta\right] \leq \sigma^2(Y/n)/\eta^2 = \frac{x(1 - x)}{n\eta^2}$$
$$\leq \frac{1}{4n\eta^2}. \qquad (8)$$

Set $\eta = \frac{1}{2}\delta(\varepsilon/2)$. Let $M/2$ be an absolute bound for $f(x)$ on the interval $[0,1]$. Then

$$|p_n(x) - f(x)| = |E(f(Y/n) - f(x))|$$
$$\leq MP[|Y/n - x| \geq \frac{1}{2}\delta(\varepsilon/2)] + \varepsilon/2 \qquad (9)$$
$$\leq \frac{M}{n\delta^2(\varepsilon/2)} + \varepsilon/2 \leq \varepsilon$$

if $n$ is taken greater than $2M/(\varepsilon \, \delta^2(\varepsilon/2))$.

The proof of the central limit theorem is somewhat more difficult. As before $X_1, \ldots, X_n$ are assumed to be *independent, identically distributed random variables with finite second moment*. Let $m = EX$, $\sigma^2 > 0$, be the common mean and variance. *The central limit theorem states that*

$$\lim_{n \to \infty} P\left(\frac{\sqrt{n}}{\sigma}\left[\frac{S_n}{n} - m\right] \le x\right) = \Phi(x) = \int_{-\infty}^{x} \frac{1}{\sqrt{2\pi}} e^{-\frac{u^2}{2}} \, du. \quad (10)$$

It is not surprising that the proof is more difficult. This result tells us much more than the law of large numbers since it indicates the rate at which the probability mass of the distribution of $S_n/n$ concentrates about the mean value $m$ as $n \to \infty$. It is enough to prove the theorem for random variables with mean zero and variance one since

$$\frac{S_n - nm}{\sigma} = \sum_{j=1}^{n} (X_j - m)/\sigma \quad (11)$$

is a sum of independent identically distributed random variables

$$(X_j - m)/\sigma \quad (12)$$

with mean zero and variance one.

The proof of the central limit theorem given is due to Petrovsky and Kolmogorov (see [40]). Let the $X_i$, $i = 1, \ldots, n$, be independent identically distributed random variables with mean zero and variance one. Let

$$\begin{aligned} p_j &= P(X = a_j) \qquad j = 1, 2, \ldots \\ \Sigma p_j &= 1 \end{aligned} \quad (13)$$

so that the $a_j$'s are the points on which the probability mass of the $X_i$'s are located. Now

$$P(X \le x) = \sum_{a_i \le x} p_i = F(x) \quad (14)$$

is a nondecreasing function of $x$ called *the distribution function of the random variable X*. Let us list the properties of a distribution function $F(x)$. We have just noted that $F(x)$ *is nondecreasing.* Further

$$\begin{aligned} \lim_{x \to -\infty} F(x) &= \lim_{x \to -\infty} P(X \le x) = 0 \\ \lim_{x \to \infty} F(x) &= P(X < \infty) = 1. \end{aligned} \quad (15)$$

The *distribution functions are* also *continuous to the right,* that is,

$$\lim_{y \to x+} F(y) = F(x + 0) = F(x) \quad (16)$$

where $y \to x+$ indicates that $y$ approaches $x$ from the right. Thus, distribution functions $F$ are nondecreasing functions with total increase one and $\lim\limits_{x \to -\infty} F(x) = 0$. The distribution functions we consider are jump functions (they increase only by jumps) since they correspond to random variables and only discrete valued random variables have been considered thus far. However, we will call any function satisfying the above conditions a distribution function even though it does not correspond to a discrete valued random variable. Later it will be shown that such functions can be made to correspond to random variables with a continuous (not necessarily discrete) value range. The reason for introducing such an enlarged notion of distribution function now is due to the fact that we have to deal with $\Phi(x)$ which is a distribution function in this enlarged sense but not in the original restricted sense. Since the mean and second moment of a discrete valued random variable $X$ are given in terms of its distribution function as

$$\int \xi \, dF, \quad \int \xi^2 \, dF \tag{17}$$

respectively, we shall generally refer to these as the mean and second moment (variance if the mean is zero) of the distribution function $F$. Thus the mean and variance of $\Phi(x)$ are zero and one respectively.

The distribution function $F_n(x)$ of $X/\sqrt{n}$ is given by

$$F_n(x) = P(X/\sqrt{n} \leq x) = F(\sqrt{n}\, x). \tag{18}$$

Let $U_{k,n}(x)$ be the distribution function of

$$\sum_{j=1}^{k} X_j/\sqrt{n}. \tag{19}$$

Now

$$
\begin{aligned}
U_{k,n}(x) &= P\left(\sum_{j=1}^{k} X_j/\sqrt{n} \leq x\right) \\
&= \sum_{i} P\left(\sum_{j=1}^{k-1} X_j/\sqrt{n} \leq x - \frac{a_i}{\sqrt{n}}, \, X_k = a_i\right) \\
&= \sum_{i} P\left(\sum_{j=1}^{k-1} X_j/\sqrt{n} \leq x - \frac{a_i}{\sqrt{n}}\right) P(X_k = a_i) \\
&= \sum_{i} U_{k-1,n}\left(x - \frac{a_i}{\sqrt{n}}\right) p_i \\
&= \int U_{k-1,n}(x - \xi) \, dF_n(\xi)
\end{aligned}
\tag{20}
$$

for $1 < k \leq n$. Now $U_n(x) = U_{n,n}(x)$ is the distribution function of

$$\sum_{j=1}^{n} X_j / \sqrt{n} \tag{21}$$

and our object is to show that

$$\lim_{n \to \infty} U_n(x) = \Phi(x). \tag{22}$$

Notice that $\Phi(x/\sqrt{t})$ is a solution of the "heat equation" or "diffusion equation"

$$\frac{\partial \Phi}{\partial t} = \frac{1}{2} \frac{\partial^2 \Phi}{\partial x^2} \tag{23}$$

in the half-plane $t > 0$. The "upper" function

$$V(x,t) = \Phi(x/\sqrt{t}) + \varepsilon\, t \tag{24}$$

($\varepsilon > 0$ a fixed positive number) plays a basic role in the proof. The function $V$ satisfies the equation

$$\frac{\partial V}{\partial t} = \frac{1}{2} \frac{\partial^2 V}{\partial x^2} + \varepsilon. \tag{25}$$

Two intermediate results or lemmas will be required. The basic idea of the proof is to replace each of the $n$ distribution functions $F_n(x)$ by the distribution function $\Phi(\sqrt{n}\, x)$. The object is to show that the error made in each such replacement is small enough so that the over-all error made is negligible. The lemmas are required in getting sufficiently good estimates of the error.

*Lemma 1: Given any $\delta > 0$ there is an $n$ (depending on $\delta$, $\varepsilon$) sufficiently large so that*

$$V\left(x, t + \frac{1}{n}\right) > \int V(x - \xi, t)\, dF_n(\xi) \tag{26}$$

*in the whole half-plane $t > \delta$.*

Now

$$V(x - \xi, t) = V(x,t) - \xi \frac{\partial V}{\partial x} + \frac{1}{2} \xi^2 \frac{\partial^2 V}{\partial x^2} + \rho(x,\xi,t) \tag{27}$$

where

$$\rho(x;\xi,t) = \frac{1}{2} \xi^2 \left[ \frac{\partial^2 V}{\partial x^2} (x - \theta\xi, t) - \frac{\partial^2 V}{\partial x^2} (x,t) \right], \; 0 < \theta < 1, \tag{28}$$

by the law of the mean. Since

$$\int dF_n(\xi) = 1, \frac{m}{\sqrt{n}} = \int \xi\, dF_n(\xi) = 0, \frac{\sigma^2}{n} = \frac{1}{n} = \int \xi^2\, dF_n(\xi) \tag{29}$$

we have

$$\int V(x - \xi, t)\, dF_n(\xi) = V(x,t) + \frac{1}{2n}\frac{\partial^2 V}{\partial x^2} + J \tag{30}$$

where

$$J = \int \rho(x,\xi,t)\, dF_n(\xi). \tag{31}$$

Now

$$|\rho(x,\xi,t)| < \xi^2/\delta \tag{32}$$

for $t > \delta$ since the absolute value of $\partial^2 V/\partial x^2$ is bounded by $1/2\delta$ in the half-plane $t > \delta$. On the other hand

$$|\rho(x,\xi,t)| < |\xi|^3/\delta^{3/2} \tag{33}$$

when $t > \delta$ by a corresponding bound on $\partial^3 V/\partial x^3$. Making use of the last inequality, we see that

$$|\rho(x,\xi,t)| < \frac{\varepsilon}{3}\xi^2 \tag{34}$$

in $t > \delta$ when $|\xi| \leq \tau = \frac{\varepsilon}{3}\delta^{3/2}$. It then follows that

$$\begin{aligned}
|J| &\leq \int_{|\xi|\leq\tau} |\rho(x,\xi,t)|\, dF_n(\xi) + \int_{|\xi|>\tau} |\rho(x,\xi,t)|\, dF_n(\xi) \\
&\leq \frac{\varepsilon}{3}\int_{|\xi|\leq\tau} \xi^2\, dF_n(\xi) + \frac{1}{\delta}\int_{|\xi|>\tau} \xi^2\, dF_n(\xi) \\
&\leq \frac{\varepsilon}{3}\frac{1}{n} + \lambda\frac{1}{\delta n}
\end{aligned} \tag{35}$$

where

$$\lambda = n\int_{|\xi|>\tau} \xi^2\, dF_n(\xi) = \int_{|\xi|>\sqrt{n}\,\tau} \xi^2\, dF(\xi). \tag{36}$$

For $n$ sufficiently large $\lambda < \varepsilon\,\delta/3$ and hence

$$|J| < \frac{2}{3}\frac{\varepsilon}{n}.$$

Since $V(x,t)$ satisfies equation (25) it follows that

$$\int V(x - \xi, t)\, dF_n(\xi) < V(x,t) + \frac{1}{n}\frac{\partial V}{\partial t} - \frac{\varepsilon}{3}\frac{1}{n}. \tag{37}$$

The relation

$$V\left(x, t + \frac{1}{n}\right) = V(x,t) + \frac{1}{n}\frac{\partial V}{\partial t} + \frac{1}{2n^2}\left[\frac{\partial^2 V}{\partial t^2}\right]_{x,t+\frac{\theta}{n}}, \quad 0 < \theta < 1, \tag{38}$$

implies that

$$V\left(x,\, t + \frac{1}{n}\right) > V(x,t) + \frac{1}{n}\frac{\partial V}{\partial t} - \frac{1}{2n^2\delta^2} \tag{39}$$

since $\left|\dfrac{\partial^2 V}{\partial t^2}\right| < \dfrac{1}{\delta^2}$ when $t > \delta$. For sufficiently large $n$ therefore

$$V\left(x,\, t + \frac{1}{n}\right) > V(x,t) + \frac{1}{n}\frac{\partial V}{\partial t} - \frac{\varepsilon}{3}\frac{1}{n}. \tag{40}$$

Lemma 1 follows from relations (37) and (40).

*Lemma 2: Let $G_1$, $G_2$ be two distribution functions with zero mean and variances less than $\beta$. Then*

$$G_1(x) - G_2(x + 2\alpha) \leq \beta/\alpha^2 \tag{41}$$

*for all $x$ and all $\alpha > 0$.*

This lemma follows readily by considering two cases and an application of Chebyshev's inequality. If $x \leq -\alpha$

$$G_1(x) \leq G_1(-\alpha) \leq \beta/\alpha^2 \tag{42}$$

and hence

$$G_1(x) - G_2(x + 2\alpha) \leq \beta/\alpha^2. \tag{43}$$

If $x > -\alpha$

$$G_2(x + 2\alpha) \geq G_2(\alpha) \geq 1 - \beta/\alpha^2 \tag{44}$$

and we again have

$$G_1(x) - G_2(x + 2\alpha) \leq 1 - G_2(x + 2\alpha) \leq \beta/\alpha^2. \tag{45}$$

The two lemmas are now applied to complete the proof of the central limit theorem. Take $\delta$ a fixed number, $0 < \delta < 1$. For some value of $s$, $s = 1, \ldots, n$

$$\delta < s/n < 2\delta.$$

The distribution function $U_{s,n}(x)$ has mean zero and variance $\dfrac{s}{n} < 2\delta$.

Now $\Phi\left(x \Big/ \sqrt{\dfrac{s}{n}}\right)$ has the same property. By Lemma 2 for all $x$ and $\alpha > 0$

$$U_{s,n}(x) - \Phi\left(\frac{x + 2\alpha}{\sqrt{\dfrac{s}{n}}}\right) < \frac{2\delta}{\alpha^2} \tag{46}$$

and thus

$$U_{s,n}(x) - V\left(x + 2\alpha, \frac{s}{n}\right) < \frac{2\delta}{\alpha^2}. \tag{47}$$

By Lemma 1, since $\frac{s}{n} > \delta$,

$$V\left(x + 2\alpha, \frac{k}{n}\right) > \int V\left(x + 2\alpha - \xi, \frac{k-1}{n}\right) dF_n(\xi) \tag{48}$$

for $k > s$. Set

$$W_k(x) = U_{k,n}(x) - V\left(x + 2\alpha, \frac{k}{n}\right). \tag{49}$$

Using (20) and (48), we obtain

$$W_k(x) < \int W_{k-1}(x - \xi) \, dF_n(\xi). \tag{50}$$

Let $\mu_k$ be the least upper bound of $W_k(x)$. Since $\int dF_n = 1$, $\mu_k \leq \mu_{k-1}(k > s)$ and hence $\mu_n \leq \mu_s$. Thus

$$U_n(x) - V(x + 2\alpha, 1) = U_n(x) - \Phi(x + 2\alpha) - \varepsilon \leq \mu_s < \frac{2\delta}{\alpha^2} \tag{51}$$

or

$$
\begin{aligned}
U_n(x) &< \Phi(x) + \frac{1}{\sqrt{2\pi}} \int_x^{x+2\alpha} e^{-\frac{1}{2}u^2} \, du + \varepsilon + \frac{2\delta}{\alpha^2} \\
&< \Phi(x) + \frac{2\alpha}{\sqrt{2\pi}} + \varepsilon + \frac{2\delta}{\alpha^2}.
\end{aligned} \tag{52}
$$

With an appropriate choice of $\alpha, \delta$

$$U_n(x) < \Phi(x) + 2\,\varepsilon. \tag{53}$$

A completely analogous argument with the "lower" function $\Phi(x/\sqrt{t}) - \varepsilon t$ leads us to

$$U_n(x) > \Phi(x) - 2\,\varepsilon \tag{54}$$

for sufficiently large $n$. Since $\varepsilon$ is an arbitrary positive number, the proof of the central limit theorem is complete.

The central limit theorem is often invoked in the theory of errors. Assume that a series of independent experiments to measure the physical constant $m$ is to be set up. The random variables $X_1, \ldots, X_n$ are the measurements of the constant $m$ in the $n$ experiments. There will generally be an error $X_i - m$ in the $i$-th experiment due to reading error, imperfections in the measuring instrument, and other such effects. Assuming not too much mass in the tail of the probability distribution of the $X$'s (existence of a second moment), it is reasonable to take the mean of the observations $\frac{1}{n} \sum_{1}^{n} X_i$ as an estimate of the physical constant $m$. Of course, it is assumed that the experiments are not biased, that is, the mean of the probability distribution of $X$ is equal to $m$. Then the central limit theorem provides an approximation for the probability distribution of the sample mean for $n$ large.

## f. Entropy of an Experiment

Consider an experiment $\mathcal{A}$ with a finite number of elementary outcomes $A_1, \ldots, A_n$ and corresponding probabilities of occurrence $p_i > 0, i = 1, \ldots, n, \Sigma p_i = 1$. We should like to associate a number $H(\mathcal{A})$ with the experiment $\mathcal{A}$ that will be a reasonable measure of the uncertainty associated with $\mathcal{A}$. Notice that $H(\mathcal{A})$ could alternatively be written as a function of the $n$ probabilities $p_i$, $H(p_1, \ldots, p_n)$, when $\mathcal{A}$ has $n$ elementary outcomes. We shall call the number $H(\mathcal{A})$ the entropy of the experiment $\mathcal{A}$.

Suppose two experiments $\mathcal{A}$, $\mathcal{B}$ with elementary outcomes $A_i$, $i = 1, \ldots, n, B_j, j = 1, \ldots, m$ respectively are considered jointly. Assume that the form of $H(\mathcal{A})$ as a function of the probabilities $p_i$ is known. It is then natural to take the conditional entropy of the experiment $\mathcal{A}$ given outcome $B_j$ of experiment $\mathcal{B}$, $H(\mathcal{A}|B_j)$, as the function of the conditional probabilities $P(A_i|B_j) i = 1, \ldots, n$ of the same form. The conditional entropy of the experiment $\mathcal{A}$ given the experiment $\mathcal{B}$, $H_{\mathcal{B}}(\mathcal{A})$, is naturally taken as

$$H_{\mathcal{B}}(\mathcal{A}) = \sum_{j=1}^{n} H(\mathcal{A}|B_j)P(B_j). \tag{1}$$

Let us now consider properties that it might be reasonable to require of the entropy of an experiment. As already remarked,

$$H(\mathcal{A}) = H(p_1, \ldots, p_n)$$

is a function of the probabilities $p_i$ of the elementary outcomes of $\mathcal{A}$. Our first assumption is that $H(p_1, \ldots, p_n)$ *is a continuous and symmetric function of* $p_1, \ldots, p_n$. Further, one feels that $H(p_1, \ldots, p_n)$ *should take its largest value when* $p_1 = \cdots = p_n = \dfrac{1}{n}$ *since this corre-*

sponds to the experiment $\mathfrak{A}$ with $n$ elementary outcomes that has the highest degree of randomness. The next property to be required is an additivity property. Let $\mathfrak{A}$, $\mathfrak{B}$ be two experiments with a finite number of elementary outcomes $A_i$, $B_j$. Let $\mathfrak{B} \vee \mathfrak{A}$ denote the joint experiment with elementary outcomes $B_i A_j$ and $H(\mathfrak{B} \vee \mathfrak{A})$ the entropy of that joint experiment. *We ask that the entropy of $\mathfrak{A}$ and $\mathfrak{B}$ jointly be equal to the sum of the entropy of $\mathfrak{A}$ and the conditional entropy of $\mathfrak{B}$ given $\mathfrak{A}$*

$$H(\mathfrak{B} \vee \mathfrak{A}) = H(\mathfrak{A}) + H_{\mathfrak{A}}(\mathfrak{B}). \tag{2}$$

The last condition is a consistency condition, namely,

$$H(p_1, \ldots, p_n) = H(p_1, \ldots, p_n, 0). \tag{3}$$

Here, the entropies of two experiments, one with $n$ outcomes and the other with $n + 1$ outcomes, have been equated. However, it is clear they are the same experiment since the $n + 1^{st}$ outcome has probability zero.

We first consider $L(n) = H\left(\dfrac{1}{n}, \ldots, \dfrac{1}{n}\right)$ and show that $L(n) = \lambda$

log $n$ *with* $\lambda$ *a positive constant*. By the second and fourth assumptions

$$\begin{aligned} L(n) &= H\left(\frac{1}{n}, \ldots, \frac{1}{n}\right) = H\left(\frac{1}{n}, \ldots, \frac{1}{n}, 0\right) \\ &\leq H\left(\frac{1}{n+1}, \ldots, \frac{1}{n+1}\right) = L(n+1). \end{aligned} \tag{4}$$

Thus $L(n)$ is a nondecreasing function of $n$. Let $m$, $r$ be positive integers. Take $m$ mutually independent experiments $S_1, \ldots, S_m$ each with $r$ equally likely elementary outcomes so that $H(S_k) = H\left(\dfrac{1}{r}, \ldots, \dfrac{1}{r}\right)$ $= L(r)$, $k = 1, \ldots, m$. The additivity of the entropy function implies that

$$H(S_1 \vee \cdots \vee S_m) = \sum_1^m H(S_k) = mL(r). \tag{5}$$

But $S_1 \vee \cdots \vee S_m$ has $r^m$ elementary equally likely events and therefore

$$H(S_1 \vee \cdots \vee S_m) = L(r^m) = mL(r). \tag{6}$$

Assume that the function $L$ is not identically zero. Consider arbitrary fixed integers $s$, $n > 0$. Take $r$ an integer greater than one with $L(r) \neq 0$.

Then an integer $m$ can be determined such that

$$r^m \leq s^n \leq r^{m+1}$$

$$m \log r \leq n \log s \leq (m + 1) \log r \tag{7}$$

$$\frac{m}{n} \leq \frac{\log s}{\log r} \leq \frac{m + 1}{n}.$$

By the monotonicity of $L$ we have

$$L(r^m) \leq L(s^n) \leq L(r^{m+1})$$

$$mL(r) \leq nL(s) \leq (m + 1)\, L(r) \tag{8}$$

$$\frac{m}{n} \leq \frac{L(s)}{L(r)} \leq \frac{m + 1}{n}.$$

Therefore

$$\left| \frac{L(s)}{L(r)} - \frac{\log s}{\log r} \right| \leq \frac{1}{n}. \tag{9}$$

On letting $n \to \infty$ we see that

$$\frac{L(s)}{L(r)} = \frac{\log s}{\log r}. \tag{10}$$

But this implies that $L(n) = \lambda \log n$ with $\lambda > 0$ because $L$ is non-decreasing.

The form of $H$ has been obtained when $p_1 = \cdots = p_n = \frac{1}{n}$. Let us now consider the form of $H$ for rational $p_i > 0$, $\Sigma p_i = 1$. Take $p_i = g_i/g$, $i = 1, \ldots, n$, $g_i > 0$, $\sum_{1}^{n} g_i = g$ with the $g_i$ integers. Consider an experiment $\mathcal{B}$ with $g$ equally likely elementary outcomes $B_1, \ldots, B_g$. Let $\mathcal{A}$ be the cruder experiment with $n$ elementary outcomes $A_1, \ldots, A_n$

$$A_1 = \bigcup_{1}^{g_1} B_i, \ldots, A_n = \bigcup_{g_1 + \cdots + g_{n-1} + 1}^{g_1 + \cdots + g_n} B_i. \tag{11}$$

Notice that $P(A_k) = g_k/g = p_k$ and the conditional probability of an event $B_i$ given $A_k$ is $1/g_k$ if $B_i$ is a subset of $A_k$ and zero otherwise. Therefore

$$\begin{aligned}
H_{\mathcal{A}}(\mathcal{B}) &= \sum_{k=1}^{n} p_k H(\mathcal{B}|A_k) = \lambda \sum_{k=1}^{n} p_k \log g_k \\
&= \lambda \sum_{k=1}^{n} p_k \log g p_k = \lambda \log g + \lambda \sum_{k=1}^{n} p_k \log p_k.
\end{aligned} \tag{12}$$

Now the experiment $\mathcal{B} \vee \mathcal{C}$ has $g$ elementary outcomes, each occurring with probability $1/g$, the remaining elementary outcomes all having probability zero. Thus $H(\mathcal{B} \vee \mathcal{C}) = \lambda \log g$. Using the additivity property of the entropy

$$H(\mathcal{C}) = H(\mathcal{B} \vee \mathcal{C}) - H_{\mathcal{C}}(\mathcal{B}) \qquad (13)$$

we find that

$$H(\mathcal{C}) = -\lambda \sum_{k=1}^{n} p_k \log p_k. \qquad (14)$$

Since the entropy $H(p_1, \ldots, p_n)$ is assumed to be a continuous function of $p_1, \ldots, p_n$, this representation is valid for real $p_1, \ldots, p_n$.

## g. Problems

1. Show that $\overline{\bigcap_{i=1}^{n} A_i} = \bigcup_{i=1}^{n} \bar{A}_i$

   and $\overline{\bigcup_{i=1}^{n} A_i} = \bigcap_{i=1}^{n} \bar{A}_i$.

   Consider the two results above for infinite collections of sets.

2. Given $A_1$ and $A_2$ show that

$$1 - P(\bar{A}_1) - P(\bar{A}_2) \leq P(A_1 A_2) \leq 1$$

   and

$$P(A_1 A_2) = 1 - P(\bar{A}_1) - P(\bar{A}_2) + P(\bar{A}_1 \bar{A}_2).$$

   Extend the results given above to collections of more than two sets.

3. Derive $P(A_1|\bar{A}_2) = P(A_1)$, $P(\bar{A}_1|A_2) = P(\bar{A}_1)$ from $P(A_1|A_2) = P(A_1)$.

4. Consider an experiment with $k$ disjoint possible outcomes where $p_1, \ldots, p_k \geq 0$ are the probabilities of the outcomes, $\sum_{1}^{k} p_i = 1$. Suppose $n$ independent identically distributed experiments of this type are conducted. Let $X_j$, $j = 1, \ldots, k$, be the number of times outcome $j$ arose in the $n$ experiments. Find the joint probability distribution of $X_1, \ldots, X_k$.

5. Find the limit of the joint distribution required in the previous example under the restraints

$$np_1 = \lambda_1 > 0, \ldots, np_{k-1} = \lambda_{k-1} > 0$$

   as $n \to \infty$.

6. Let $X_1, \ldots, X_n$ be independent identically distributed binomial variables of sample size one each with probability $p$ of success. Show that the probability distribution of $X_1 + \cdots + X_n$ is the binomial distribution of sample size $n$.

7. Let $X$, $Y$ be independent binomial distributed random variables of sample size $n$ and $m$ respectively with the probability of a success $p$. Show that the probability distribution of $X + Y$ is binomial with sample size $n + m$.

8. Let $X$, $Y$ be independent Poisson distributed random variables with means $\lambda_1, \lambda_2 > 0$ respectively. Show that the probability distribution of $X + Y$ is Poisson with mean $\lambda_1 + \lambda_2$.

9. Let $X$ be a random variable that can assume only non-negative integer values. Let $g(s) = \sum_{k=0}^{\infty} p_k s^k$, $p_k = P(X = k)$, be the generating function of the distribution of $X$. Does knowledge of $g(s)$ determine the probability distribution of $X$? Why?

10. Let $f(t)$ be a continuous even function on $[-\pi, \pi]$. By setting $x = \cos t$ and using the Weierstrass approximation theorem, show that $f(t)$ can be uniformly approximated by finite trigonometric series in $\sin kt$, $\cos kt$. Use this result to show that any continuous function $f(t)$ on $[-\pi, \pi]$ with $f(\pi) = f(-\pi)$ can be uniformly approximated by finite trigonometric series.

11. The functions in $x$ and $y$ which are weighted sums of $\binom{n}{k}\binom{n}{j} x^k$ $(1 - x)^{n-k} y^j (1 - y)^{n-j}$, $k, j = 0, 1, \ldots, n$, are analogues of the Bernstein polynomials. Prove the Weierstrass approximation theorem for a continuous function of two variables on the unit square using these polynomials.

12. Show that $\partial^2 V / \partial x^2$ is bounded in absolute value by $1/2\delta$ in the half-plane $t > \delta > 0$ where $V(x,t) = \Phi(x/\sqrt{t}) + \varepsilon t$ and $\Phi$ is the normal distribution function given by (e.10).

13. Show that $\partial^3 V / \partial x^3$ is bounded in absolute value by $2/\delta^{3/2}$ in $t > \delta > 0$ where $V$ is the function given in Problem 12.

14. Let $X_j$, $j = 1, \ldots, n$, be independent random variables with mean zero and variance $\sigma_j^2 > 0$. The distribution function of $X_j$ is $F_j(x)$. Let $B_n = \sum_{j=1}^{n} \sigma_j^2$. Show how to modify the proof of the

central limit theorem of section e so as to obtain asymptotic normality of $\sum\limits_{j=1}^{n} X_j/\sqrt{B_n}$ under the assumption that

$$\frac{1}{B_n} \sum_{k=1}^{n} \int_{|x|>\tau\sqrt{B_n}} x^2 \, dF_k(x) \to 0$$

as $n \to \infty$ for every $\tau > 0$.

**15.** Show that if $\sum\limits_{1}^{n} E|X_j|^\alpha/B_n^{\alpha/2} \to 0$, $\alpha > 2$, as $n \to \infty$ then

$$\sum_{1}^{n} \int_{|x|>\tau B_n^{1/2}} x^2 \, dF_k(x)/B_n \to 0$$

for any fixed $\tau > 0$ where $B_n$ is as defined in the previous example. This remark coupled with the previous example gives us Liapounoff's form of the central limit theorem.

**16.** A function $f$ defined on $(0, \infty)$ is called convex if

$$f(\lambda x + (1 - \lambda)y) \le \lambda f(x) + (1 - \lambda)f(y)$$

for all $\lambda$, $0 \le \lambda \le 1$, and $x$, $y$ in its domain of definition. Indicate what this means geometrically. Show that if $f''(x)$ exists and is non-negative everywhere, the function $f$ is convex. Apply this to $f(x) = x \log x$.

**17.** Show that a convex function $f$ satisfies the inequality

$$f\left( \sum_{i=1}^{n} p_i x_i \right) \le \sum p_i f(x_i)$$

for all $x_1, \ldots, x_n$ and all $p_i$, $p_i \ge 0$, $\sum p_i = 1$. This inequality is called *Jensen's inequality*.

**18.** Find the entropy of (a.) the binomial distribution; (b.) the Poisson distribution.

**19.** Using Jensen's inequality show that $H_\alpha(\mathfrak{B}) \le H(\mathfrak{B})$. What happens to the inequality when $\mathfrak{A}$ and $\mathfrak{B}$ are independent experiments. Use the inequality above to obtain $H(\mathfrak{A} \vee \mathfrak{B}) \le H(\mathfrak{A}) + H(\mathfrak{B})$.

# Notes

1. Many of the basic ideas and tools of probability theory are discussed in sections a, b, and d for probability spaces with at most a countable number of points. They are introduced in Chapter IV again for general probability spaces. One hopes that the treatment given for the discrete case in this chapter will intuitively motivate the discussion in Chapter IV for general probability spaces.

2. The binomial and Poisson distributions of section c are the discrete probability distributions that most commonly arise in theory and practice. An extensive discussion of other discrete distributions and of a wide range of combinatorial problems in which they arise can be found in Feller [15].

3. A derivation of the Weierstrass approximation theorem ordinarily would not be given in a text on probability theory. It is given in section e because it follows immediately from the law of large numbers by S. Bernstein's simple and beautiful proof. Notice that it explicitly produces an approximation of simple form. Further, a more detailed analysis would give bounds on the error of approximation in terms of the regularity of the continuous function approximated.

4. The proof of the central limit theorem given in section e is an older proof but it is well worth reviving for expository purposes. Most of the current proofs have limited intuitive appeal because they use a circuitous argument via some version of transform theory. The derivation presented in this chapter certainly does not have this failing. Further, even though some detailed estimates are required, the proof is elementary in character. An interesting recent paper of C. Stein [A15] also uses a direct approach to estimate the error in the normal approximation to the distribution of a sum of dependent random variables.

# MARKOV CHAINS

## a. The Markov Assumption

Thus far we have discussed models of independent observations. In fact, the most detailed and classical investigations in Probability Theory are centered about the notion of independence. However, there are many contexts which require models in which some notion of statistical dependence is basic. The simplest models of this sort are based on the Markov assumption. Even though this assumption does not appear to allow radical departures from independence, we shall later on see that the study of such Markov schemes or processes will give us great insight in studying various types of statistical dependence.

Consider a system that is to be studied at discrete time points $t = 1, \ldots, n$ and whose possible states at each time $t$ can be completely labeled by the integers $i = 1, 2, \ldots$. It is convenient to refer to the possible states at a fixed time as the *state space* of the model. The various possible histories of the system or *sample points* of the probability space to be constructed are given by $n$-vectors of integers $w = (i_1, \ldots, i_n)$. Assume that we are given a vector

$$\mathbf{p} = (p_1, p_2, \ldots) \tag{1}$$

of non-negative numbers $p_i$, $\sum_1^\infty p_i = 1$ and matrices

$$\mathbf{P}^{(m,m+1)} = (p_{i_m, i_{m+1}}^{(m,m+1)}; i_m, i_{m+1} = 1, 2, \ldots), \tag{2}$$
$$m = 1, \ldots, n - 1,$$

with non-negative elements and row sums one

$$\sum_{i_{m+1}=1}^\infty p_{i_m, i_{m+1}}^{(m,m+1)} = 1. \tag{3}$$

The Markov assumption states that the probability of a sample point $(i_1, \ldots, i_n)$ is given by the product

$$P((i_1, \ldots, i_n)) = p_{i_1} p_{i_1, i_2}^{(1,2)} \cdots p_{i_{n-1}, i_n}^{(n-1,n)}. \tag{4}$$

The $m$-th coordinate $X_m(w) = i_m$, $m = 1, \ldots , n$, of a sample point $w$ is a *random variable* and represents the state of the system at time $m$. Such a family of random variables on a probability space is an example of a random or stochastic process. Making use of formula (4), it is now clear that the vector $\mathbf{p}$ is the vector of initial probabilities

$$p_i = P(X_1(w) = i) \tag{5}$$

and the elements of the matrices $\mathbf{P}^{(m,m+1)}$ are one-step conditional probabilities

$$p_{i_m,i_{m+1}}^{(m,m+1)} = P(X_{m+1} = i_{m+1}|X_m = i_m), \tag{6}$$
$$m = i, \ldots , n - 1.$$

Such a probability model is called a *Markov chain*. Thus, the Markov assumption states that joint probabilities can be computed in a simple manner (as given by (4)) from an initial probability distribution and one-step transition probabilities or conditional probabilities. The simplest and most carefully studied case is that in which the transition probability matrices $\mathbf{P}^{(m,m+1)}$ are independent of $m$. We shall refer to this as the case of *stationary transition mechanism* and discuss it in some detail later on. If all the one-step transition probabilities

$$p_{i_m,i_{m+1}}^{(m,m+1)} \tag{7}$$

do not depend on the initial subscript $i_m$, the random variables $X_m$, $m = 1, \ldots , n$, are independent random variables.

The Markov assumption can be recast in another form that gives greater insight into its intuitive meaning. Consider time $t = m$ as the present. Let $A$ be any set of sample points obtained by restrictions on the possible states of the system in the past, that is, $t < m$, and $B$ any set of sample points obtained by restrictions on the possible states of the system in the future, $t > m$. Thus, we would have $A, B$ of the form

$$A = \{a_1 \leq X_1 \leq b_1, \ldots , a_{m-1} \leq X_{m-1} \leq b_{m-1}\}$$
$$B = \{a_{m+1} \leq X_{m+1} \leq b_{m+1}, \ldots , a_n \leq X_n \leq b_n\}. \tag{8}$$

Suppose it is known that the system is in the state $i_m$ at time $m$. Then the Markov assumption (4) and the additivity of the probability function indicates that the conditional probability of the joint occurrence of $A$ and $B$ factors

$$P(AB|X_m = i_m) = P(A|X_m = i_m)P(B|X_m = i_m). \tag{9}$$

Thus, *given precise knowledge of the present, the past (A) and future (B) are independent*. We have only shown that the Markov assumption implies this property. However, one can easily show that this reformulation of the Markov property implies (4).

Higher order transition probabilities can be easily computed in terms of one-step transition probabilities

$$P(X_{m+r} = i_{m+r} \mid X_m = i_m) = p^{(m,m+r)}_{i_m,i_{m+r}} =$$
$$= \sum_{\substack{i_{m+1} \\ \vdots \\ i_{m+r-1}}} p^{(m,m+1)}_{i_m,i_{m+1}} p^{(m+1,m+2)}_{i_{m+1},i_{m+2}} \cdots p^{(m+r-1,m+r)}_{i_{m+r-1},i_{m+r}}. \tag{10}$$

Call the matrix of transition probabilities $p^{(m,m+r)}_{i_m,i_{m+r}}$, $\mathbf{P}^{(m,m+r)}$. A matrix equation for computing higher order transition probabilities from those of lower order follows readily

$$\mathbf{P}^{(r,s)}\mathbf{P}^{(s,t)} = \mathbf{P}^{(r,t)}, \, r < s < t. \tag{11}$$

This equation is usually called the *Chapman-Kolmogorov equation*. The equation reduces simply in the case of a Markov chain with stationary transition mechanism for then

$$\mathbf{P}^{(r,s)} = \mathbf{P}^{s-r}, \, s > r, \tag{12}$$

where $\mathbf{P} = \mathbf{P}^{(m,m+1)}$, $m = 1, \ldots, n-1$. The higher order transition probability matrices are just powers of the one-step transition probability matrix. All the relations obtained above have been written out for a Markov chain with a denumerable infinity of states. All the relations are the same, of course, in the case of a chain with a finite number of states. They differ only in that infinite sums are replaced by finite sums.

A simple example of a Markov chain without stationary transition mechanism is given by the chain with

$$p^{(m,m+1)}_{i,j} = \begin{cases} \dfrac{\lambda_m^{j-i}}{(j-i)!} e^{-\lambda_m} & \text{if } j \geq i \\ 0 & \text{otherwise} \end{cases} \tag{13}$$

$i, j = 1, 2, \ldots,$ and $m = 1, \ldots, n-1$, where $\lambda_1 \neq \lambda_2 \neq \cdots \neq \lambda_{n-1}$, $\lambda_i > 0$. Of course, if we allow $\lambda_1 = \lambda_2 = \cdots = \lambda_{n-1}$ a chain with stationary transition mechanism is obtained. Since the row distributions of the matrices $\mathbf{P}^{(m,m+1)}$ are Poisson, it follows that the elements $p_{i,j}^{(m,m+r)}$ of $\mathbf{P}^{(m,m+r)}$ are given by

$$p_{i,j}^{(m,m+r)} = \begin{cases} \dfrac{(\sum\limits_{i=m}^{m+r-1} \lambda_i)^{j-i}}{(j-i)!} \, e^{-\sum\limits_{i=m}^{m+r-1} \lambda_i} & \text{if } j \geq i, \\ 0 & \text{otherwise,} \end{cases} \tag{14}$$

$r > 0$. This Markov chain could be taken as a model of a telephone exchange as observed at the discrete time points $t = 1, \ldots, n$. The random variable $X_j$ is the number of calls made through the exchange from time $t = 1$ through time $t = j$. The additional number of calls made in the time interval $j < t \leq j + 1$ is assumed to be governed by a Poisson distribution with mean $\lambda_j$ and independent of the calls already made. Such a Markov chain is sometimes called a *growth* process because the probability mass drifts into the states with larger index as time goes. This is obvious since the number of calls increases as time goes on.

The next example is sometimes taken as a model of population growth or death. Assume that we are studying a homogeneous population whose growth (and death) mechanism does not change with time. $X_s$ denotes the number of individuals in the population at time $s$. The probability of one individual at time $s$ generating $j$ individuals at time $s + 1$ is given by $q_j \geq 0$, where

$$\sum_{j=0}^{\infty} q_j = 1. \tag{15}$$

If there are $i$ individuals at time $s$, they are assumed to act independently of each other in generating progeny for the next generation. Thus, $p_{i,j}$, the probability of $i$ individuals at time $s$ generating $j$ progeny at time $s + 1$, is given by

$$p_{i,j} = \sum_{j_1 + \cdots + j_i = j} q_{j_1} q_{j_2} \cdots q_{j_i} = q_j^{(i*)}, \tag{16}$$

the $j$-th element in the vector obtained by convoluting the sequence $q = (q_0, q_1, \ldots)$ with itself $i$ times. Let $\varphi(s)$ be the generating function of the $q$ sequence, that is,

$$\varphi(s) = \sum_{j=0}^{\infty} q_j s^j. \tag{17}$$

It is clear that $q_j^{(i*)}$ is the coefficient of $s^j$ in the power series expansion of $\varphi(s)^i$.

Suppose we start with one individual at time $t = 0$. The probability of $j$ individuals at time $t = 1$ is $q_j$, the coefficient of $s^j$ in the power series expansion of $\varphi(s)$. Let us compute the probability of a population of $j$ individuals at time $t = 2$. Suppose there are $i$ individuals at time $t = 1$. The joint probability of $i$ individuals at $t = 1$ and $j$ at time $t = 2$ is given by $q_i q_j^{(i*)}$. The probability of $j$ at time $t = 2$ is obtained by summing over $i$

$$\sum_{i=0}^{\infty} q_i q_j^{(i*)}. \tag{18}$$

But this is the coefficient of $s^j$ in the power series expansion of

$$\varphi^{(2)}(s) = \varphi(\varphi(s)) = \sum_{i=0}^{\infty} q_i \varphi(s)^i. \tag{19}$$

Essentially the same argument indicates that the probability of $j$ individuals at time $t + 1$ is given by the coefficient of $s^j$ in the expansion of

$$\varphi^{(t+1)}(s) = \varphi(\varphi^{(t)}(s)) \quad t = 1, 2, \ldots . \tag{20}$$

Certain aspects of the growth or death of the population can be described by studying the iterates $\varphi^{(t)}(s)$ of the generating function $\varphi(s)$.

Let $e_t$ be the probability that the population dies out before or at time $t$, that is,

$$e_t = P(X_t = 0). \tag{21}$$

Thus $e_t = \varphi^{(t)}(0)$. Suppose we are interested in the probability of eventual extinction of the population, namely, $\lim_{t \to \infty} e_t = e$. This limit exists since $e_t$ is a nondecreasing bounded sequence. Note that $\varphi(s)$ is a continuous nondecreasing non-negative function on $0 \leq s \leq 1$ with $\varphi(1) = 1$. It is clear that $e$ is a solution of the equation

$$e = \varphi(e) \tag{22}$$

for

$$e = \lim_{t \to \infty} e_{t+1} = \lim_{t \to \infty} \varphi(\varphi^{(t)}(0)) = \varphi(e). \tag{23}$$

In fact $e$ is the smallest solution of this equation in the interval $[0,1]$ (see figure 3.1). For if $\zeta$ is any other solution in $[0,1]$

$$e = \lim_{t \to \infty} \varphi^{(t)}(0) \leq \lim_{t \to \infty} \varphi^{(t)}(\zeta) = \zeta \tag{24}$$

since $\varphi$ is nondecreasing. Consider the solutions of $\zeta = \varphi(\zeta)$ in $[0,1]$.

The value $\zeta = 1$ is one such solution. If there is another solution $\zeta$, $0 \leq \zeta < 1$,

$$\frac{\varphi(1) - \varphi(\zeta)}{1 - \zeta} = 1. \tag{25}$$

By the mean value theorem there is a point $\sigma$, $\zeta < \sigma < 1$, at which $\varphi'(\sigma) = 1$. Now $\varphi'(s) = \sum\limits_{k=0}^{\infty} k q_k s^{k-1}$ is a continuous nondecreasing function of $s$, $0 \leq s < 1$. Thus, if there is such a point $\sigma$, $\varphi'(1) \geq 1$.

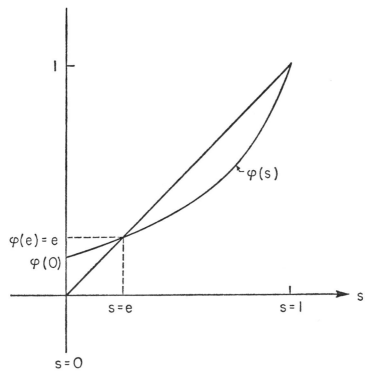

FIG. 3.1. Graph of the generating function $\varphi(s)$.

In fact, excluding the trivial case $\varphi(s) = s$, $\varphi'(s)$ is strictly increasing and $\varphi'(1) > 1$; there can be at most one such point $0 \leq \zeta < 1$ such that $\zeta = \varphi(\zeta)$. Conversely if $\varphi'(1) > 1$ there is a solution $\zeta$, $0 \leq \zeta < 1$, of (22) since $\varphi(0) \geq 0$. Notice that $\varphi'(1) = \Sigma k q_k = \mu$ is the expected number of individuals produced by one generation. We therefore have the following result. *If $\mu \leq 1$ (except for the trivial case $q_1 = 1$) the prob-*

*ability of eventual extinction of the population is one. If $\mu > 1$, there is a unique solution $\zeta$, $0 \leq \zeta < 1$, of (22) and $\zeta$ is the probability of eventual extinction of the population.* If there are many individuals, say $r$, the probability of eventual extinction when $\zeta < 1$ is small, for it is $\zeta^r$ assuming the individuals act independently.

It is very easy to compute the expected number of individuals at time $n$ for it is given by

$$EX_n = \frac{d}{ds}\,\varphi^{(n)}(s)_{|s=1} = \frac{d}{ds}\,\varphi(\varphi^{(n-1)}(s))_{|s=1}$$

$$= \varphi'(1)\frac{d}{ds}\,\varphi^{(n-1)}(s)_{|s=1} = \;\cdots\; = \mu^n. \tag{26}$$

If $\mu > 1$ the growth of the population is exponential in the mean. Such models are called branching processes. They have been used as models of bacterial colony growth and chain reactions. An intensive discussion of such models can be found in a paper of T. Harris [31]. A treatment of corresponding multi-population models can be found in another paper of Harris [32].

A third illustrative example is provided by some simple discrete models of a diffusion process. Consider a particle in a random walk on the integer points of the real line. The states are labeled by the integers $i = 0, \pm 1, \ldots$ . Given that the particle is at $i$ at time $t$, its probability of going one step to the right is given by $p$ and one step to the left by $q$, $p, q \geq 0$, $p + q = 1$. Thus

$$p_{i,i+1} = p, \qquad p_{i,i-1} = q. \tag{27}$$

Assuming that the particle is at $i = 0$ at $t = 0$, let us compute the probability that it will be at $i = j$ at time $t = n$. The particle can only reach even points in an even number of steps and odd points in an odd number of steps. For convenience, the computation will only be carried out for $j$, $n$ even. The particle can only end up at $j$ if it makes $(n + j)/2$ steps to the right and $(n - j)/2$ steps to the left. There are precisely binomial coefficient

$$\binom{n}{(n+j)/2} \tag{28}$$

such distinct paths and each of them has probability

$$p^{\frac{n+j}{2}} q^{\frac{n-j}{2}}. \tag{29}$$

The probability desired is therefore given by

$$\binom{n}{(n+j)/2} p^{\frac{n+j}{2}} q^{\frac{n-j}{2}}. \tag{30}$$

Consider now a random walk with an absorbing barrier at $-a < 0$. The states are now labeled by the integers $i = -a, -a+1, \ldots$. The transition probabilities from the states $i > -a$ are given as before

$$p_{i,i+1} = p, \qquad p_{i,i-1} = q, \tag{31}$$

$p, q \ge 0, p + q = 1$. However,

$$p_{-a,-a} = 1. \tag{32}$$

Thus, if the particle ever enters state $-a$ in its walk, it is held fast there from then on. Again assuming the particle is at $0$ at time $t = 0$, we would like to compute the probability that the particle is at $j > -a$ at time $t = n$. For convenience, assume that $j$ and $n$ are odd. In computing the desired probability the paths in the unrestricted random walk (without a barrier) that go from $0$ to $j$ in $n$ steps and pass through $-a$ must be deleted. However, there is a one-one correspondence between the paths that go from $0$ to $j$ through $-a$ in $n$ steps and those that go from $-2a$ to $j$ in $n$ steps. Consider any specific path that goes from $0$ to $j$ through $-a$. Let $t$, $0 < t < n$, be the first time the particle following the path is at $-a$. Take the mirror image of this first part of the path, from time $0$ to time $t$, with respect to $-a$. We then have a path that starts at $-2a$ at $0$ and is at $-a$ at time $t$. Leave the remainder of the path from time $t$ to time $n$ as it was. The new path is then a path from $-2a$ to $j$. A little reflection indicates that this correspondence is one-one. Thus the total number of paths from $0$ to $j$ in time $n$ that do not pass through $-a$ is equal to the total number of paths from $0$ to $j$ in the unrestricted random walk less the total number of paths from $-2a$ to $j$ in an unrestricted walk

$$\binom{n}{(n+j)/2} - \binom{n}{\frac{(n+j)}{2} + a}. \tag{33}$$

Since each path has probability $p^{\frac{n+j}{2}} q^{\frac{n-j}{2}}$, the probability of reaching $j$ in $n$ steps is

$$\left[ \binom{n}{(n+j)/2} - \binom{n}{\frac{(n+j)}{2} + a} \right] p^{\frac{n+j}{2}} q^{\frac{n-j}{2}}. \tag{34}$$

Let us now return to a brief discussion of a Markov chain with stationary transition mechanism. The transition mechanism is governed by the one-step transition probability matrix $\mathbf{P} = (p_{i,j})$. Given a transition matrix $\mathbf{P}$ one may or may not be able to find a left invariant probability vector $\mathbf{p} = (p_i)$, $p_i \geq 0$, $\Sigma p_i = 1$

$$\mathbf{p}\mathbf{P} = \mathbf{p}. \tag{35}$$

Given the existence of such a vector, it is of interest to look at the Markov chain with initial distribution $\mathbf{p}$ and transition matrix $\mathbf{P}$. The instantaneous probability distribution

$$P(X_s = i) = p_i, \quad s = 0, 1, \ldots, n \tag{36}$$

is invariant under time shift. In fact, the probability of the events $\{X_{s_1+t} = i_1, \ldots, X_{s_\alpha+t} = i_\alpha\}$, $s_1 < \cdots < s_\alpha$, is independent of $t$

$$P(X_{s_1+t} = i_1, \ldots, X_{s_\alpha+t} = i_\alpha) = p_{i_1} p_{i_1,i_2}^{(s_2-s_1)} \cdots p_{i_{\alpha-1},i_\alpha}^{(s_\alpha-s_{\alpha-1})} \tag{37}$$

($p_{i,j}^{(s)}$ is the $(i,j)$-th element of $\mathbf{P}^s$). Thus, the probability structure of the process is invariant under time shifts. Such processes are called *stationary processes*. Processes like this are reasonable as models when the probability structure of the phenomenon described is stable through time. In Chapter V we shall examine stationary processes that are more complex in structure.

Invariant probability vectors will be shown to exist for any finite dimensional transition probability matrix $\mathbf{P}$ in section b. However, they need not exist when $\mathbf{P}$ is infinite dimensional. An example illustrating this last remark is given by the matrix $\mathbf{P}$ for the growth process with

$$p_{i,j} = \begin{cases} \frac{1}{2} & \text{if } j = i, i+1 \\ 0 & \text{otherwise.} \end{cases} \tag{38}$$

A left invariant probability vector $\mathbf{p}$ would have to satisfy

$$\sum_i p_i p_{i,j} = \frac{1}{2}(p_{j-1} + p_j) = p_j \tag{39}$$

for $j > 1$ and therefore all its components would have to be equal. This is obviously impossible.

# b. Matrices with Non-negative Elements (Approach of Perron-Frobenius)

We shall discuss certain structural properties of finite square matrices $\mathbf{A} = (a_{i,j}; i, j = 1, \ldots, n)$ with non-negative elements $a_{i,j} \geq 0$ in this section. Finite probability matrices correspond to the special case obtained by setting row sums $\sum_j a_{i,j}$ equal to one. This is a condition on preservation of mass. The more general case discussed here allows for generation or destruction of mass. Our approach is non-probabilistic and due to Perron and Frobenius. The very elegant treatment given here is due to H. Wielandt [76]. In section c analogues of some of the results will be established for transition probability matrices in both finite and infinite dimensional cases by probability methods.

A number $\lambda$ is said to be a *right eigenvalue* of the matrix $\mathbf{A}$ if there is a nontrivial solution $\mathbf{x} \neq 0 = (0, \ldots, 0)$ of the equation

$$\mathbf{A}\mathbf{x}' = \lambda\mathbf{x}'.* \tag{1}$$

The column vector $\mathbf{x}'$ is a corresponding right eigenvector. Now (1) is satisfied if and only if $\mathbf{A} - \lambda\mathbf{I}$ ($\mathbf{I}$ the identity matrix) is singular. Thus $\lambda$ is also a *left eigenvalue* since there is a vector $\mathbf{y}$ such that $\mathbf{y}\mathbf{A} = \lambda\mathbf{y}$. Of course $\mathbf{x}, \mathbf{y}$ will generally be different vectors. We shall simply refer to $\lambda$ as an *eigenvalue* of $\mathbf{A}$. It is clear that the eigenvalues $\lambda$ of $\mathbf{A}$ are the solutions of the characteristic equation

$$\Phi(\lambda) = \det(\mathbf{A} - \lambda\mathbf{I}) = 0. \tag{2}$$

Knowledge of the eigenvalues $\lambda$ of $\mathbf{A}$ gives one considerable information about the behavior of the elements $a_{i,j}^{(m)}$ of $\mathbf{A}^m$ for large $m$. Consider the generating function

$$\mathbf{A}(z) = \sum_{j=0}^{\infty} z^j \mathbf{A}^j = (\mathbf{I} - z\mathbf{A})^{-1} \tag{3}$$

which is well defined for sufficiently small $|z|$. By Cramer's rule the $(i,j)$-th element $a_{i,j}(z)$ of $\mathbf{A}(z)$ is given by

$$a_{i,j}(z) = \frac{p_{i,j}(z)}{\det(\mathbf{I} - z\mathbf{A})} = \frac{p_{i,j}(z)}{(1 - \lambda_1 z) \cdots (1 - \lambda_n z)} \tag{4}$$

where $p_{i,j}(z)$ is the cofactor of the $(j,i)$-th element in $\mathbf{I} - z\mathbf{A}$ and $\lambda_1, \ldots, \lambda_n$ are the eigenvalues of $\mathbf{A}$. If the eigenvalues $\lambda_i$ are distinct we see that

$$a_{i,j}(z) = \sum_{m=0}^{\infty} a_{i,j}^{(m)} z^m$$

where

$$a_{i,j}^{(m)} = \begin{cases} \delta_{ij} & \text{if } m = 0 \\ \sum_{k=1}^{n} b_{i,j}^{(k)} \lambda_k^m & \text{otherwise,} \end{cases} \tag{5}$$

that is, $a_{i,j}^{(m)}$ is a weighted sum of the $m$-th powers of the eigenvalues.

* Given a rectangular matrix $\mathbf{A}$, $\mathbf{A}'$ is its transpose, that is $a'_{i,j} = a_{j,i}$.

An $n \times n$ matrix $\mathbf{A}$ is said to be *irreducible* if one cannot bring $\mathbf{A}$ into a form

$$\begin{pmatrix} \mathbf{A}_{11} & \mathbf{A}_{12} \\ \hline 0 & \mathbf{A}_{22} \end{pmatrix} \tag{6}$$

by a consistent relabeling of the rows and columns where $\mathbf{A}_{11}$ is an $r \times r$ matrix, $1 \leq r < n$, and $\mathbf{A}_{22}$ is an $(n - r) \times (n - r)$ matrix. Otherwise $\mathbf{A}$ is *reducible*. We shall be interested in irreducible non-negative matrices. A vector $\mathbf{x}$ will be called non-negative if all its components $x_i$ are non-negative.

**Theorem 1:** *The characteristic equation of the non-negative irreducible matrix* $\mathbf{A}$

$$\Phi(z) = \det(z\mathbf{I} - \mathbf{A}) = 0 \tag{7}$$

*has a simple positive root* $\lambda$ *that is greater than or equal to the absolute value of any other root. The eigenvector of* $\mathbf{A}$ *corresponding to this root can be taken with all its components positive. The maximal eigenvalue* $\lambda$ *is the only one with its corresponding eigenvector non-negative.*

Let $\mathbf{x}$ be any non-negative vector that is nontrivial ($\mathbf{x} \neq 0$). Set

$$\lambda_{\mathbf{x}} = \min_i \frac{\sum_j a_{i,j} x_j}{x_i} \tag{8}$$

with the convention that the fraction is set equal to $+\infty$ if $x_i = 0$. Notice that $\lambda_{\mathbf{x}}$ is the largest number for which

$$\mathbf{Ax'} - \lambda_{\mathbf{x}} \mathbf{x'} \geq 0. \tag{9}$$

Let $\mathbf{y}$ be the vector with all components equal to one. Then

$$\lambda_{\mathbf{x}} \leq \frac{\mathbf{yAx'}}{\mathbf{yx'}} \leq \frac{C\mathbf{yx'}}{\mathbf{yx'}} = C \tag{10}$$

where $C$ is the largest component of the vector $\mathbf{yA}$. Thus the numbers $\lambda_{\mathbf{x}}$ are bounded above. Let $\lambda$ be the least upper bound of the numbers $\lambda_{\mathbf{x}}$

$$\lambda = \max_{\substack{\mathbf{x} \geq 0 \\ \mathbf{x} \neq 0}} \min_i \frac{\sum_j a_{i,j} x_j}{x_i} \tag{11}$$

This upper bound $\lambda$ is positive since $\lambda_{\mathbf{x}} > 0$ for $\mathbf{x} = \mathbf{y}$ ($\mathbf{A}$ has no row consisting entirely of zeros since it is irreducible).

It is clear that there are *extremal* vectors, that is, vectors $\mathbf{z} \neq 0$, $\mathbf{z} \geq 0$ such that $\mathbf{Az}' - \lambda\mathbf{z}' \geq 0$. Note that these are vectors for which $\lambda$ is attained, namely $\lambda = \lambda_z$. Nonetheless, by the very definition of $\lambda$ there is no vector $\mathbf{x} \geq 0$ such that $\mathbf{Ax}' - \lambda\mathbf{x}' > 0$. We wish to show that *every extremal vector $\mathbf{z}$ is an eigenvector of $\mathbf{A}$ with eigenvalue $\lambda$ and that $\mathbf{z}$ is positive.* Consider any vector $\mathbf{x} \geq 0$, $\mathbf{x} \neq 0$. We shall first show that $(\mathbf{I} + \mathbf{A})^{n-1}\mathbf{x}' > 0$. Set $(\mathbf{I} + \mathbf{A})^\nu\mathbf{x}' = \mathbf{x}^{(\nu)'}$. Then $\mathbf{x}^{(\nu+1)} = \mathbf{x}^{(\nu)} + \mathbf{x}^{(\nu)}\mathbf{A}' \geq \mathbf{x}^{(\nu)} \geq 0$. Thus in $\mathbf{x}^{(\nu+1)}$ at most those components that vanished in $\mathbf{x}^{(\nu)}$ can be zero. However, it might be that exactly the same components in $\mathbf{x}^{(\nu)}$ and $\mathbf{x}^{(\nu+1)}$ vanish. But then we can write

$$\mathbf{x}^{(\nu)'} = \begin{pmatrix} \mathbf{u}' \\ 0 \end{pmatrix}, \mathbf{u} > 0$$

$$\mathbf{x}^{(\nu+1)'} = \mathbf{x}^{(\nu)'} + \mathbf{Ax}^{(\nu)'} = \begin{pmatrix} \mathbf{u}' \\ 0 \end{pmatrix} + \begin{pmatrix} \mathbf{A}_{11} & \mathbf{A}_{12} \\ \mathbf{A}_{21} & \mathbf{A}_{22} \end{pmatrix} \begin{pmatrix} \mathbf{u}' \\ 0 \end{pmatrix} \qquad (12)$$

$$= \begin{pmatrix} \mathbf{v}' \\ 0 \end{pmatrix}.$$

But then $\mathbf{A}_{21}\mathbf{u}' = 0$ implying that $\mathbf{A}_{21} = 0$ contrary to the assumption of irreducibility of $\mathbf{A}$, a contradiction. Since $\mathbf{x} = \mathbf{x}^{(0)}$ has at most $n - 1$ components zero, this implies that $\mathbf{x}^{(n-1)}$ has no zeros. Now let $\mathbf{z}$ be an extremal vector so that $\mathbf{z} \neq 0$, $\mathbf{z} \geq 0$, $\mathbf{Az}' - \lambda\mathbf{z}' = \mathbf{x}' \geq 0$. If $\mathbf{x} \neq 0$, $(\mathbf{I} + \mathbf{A})^{n-1}\mathbf{x}' = \mathbf{Ay}' - \lambda\mathbf{y}' > 0$ where $\mathbf{y}' = (\mathbf{I} + \mathbf{A})^{n-1}\mathbf{z}'$. Since this is impossible $\mathbf{x} = 0$ and $\mathbf{z}$ is an eigenvector of $\mathbf{A}$ with eigenvalue $\lambda$. Moreover $\mathbf{z} > 0$ since $0 < \mathbf{y} = (1 + \lambda)^{n-1}\mathbf{z}$.

Given any matrix $\mathbf{M} = (m_{i,j})$ let $\mathbf{M}^* = (|m_{i,j}|)$ be the matrix with elements the absolute values of the corresponding elements of $\mathbf{M}$. Now let $\alpha$ be any eigenvalue of $\mathbf{A}$ so that $\mathbf{Ax}' = \alpha\mathbf{x}'$ for some $\mathbf{x} \neq 0$. Then

$$|\alpha|\mathbf{x}^{*'} \leq \mathbf{Ax}^{*'}, |\alpha| \leq \lambda_{\mathbf{x}^*} \leq \lambda \qquad (13)$$

so that *$\lambda$ is the eigenvalue with maximal absolute value.* Further *$\lambda$ is the only eigenvalue with a non-negative eigenvector.* For suppose there were another such eigenvalue $\alpha$ with eigenvector $\mathbf{x} \geq 0$, $\mathbf{x} \neq 0$. Let $\mathbf{y}$ be an extremal vector of $\mathbf{A}'$. Then

$$\alpha\mathbf{yx}' = \mathbf{yAx}' = (\mathbf{yA})\mathbf{x}' = \lambda\mathbf{yx}'. \qquad (14)$$

Then $\alpha = \lambda$ since $\mathbf{y} > 0$ implies that $\mathbf{yx}' \neq 0$.

Let $\mathbf{x}$ be an eigenvector of $\mathbf{A}$ with eigenvalue $\lambda$ and $\mathbf{z}$ a given extremal vector. Take $c$ so that $\mathbf{x} - c\mathbf{z} = \mathbf{y} \geq 0$ and one component of $\mathbf{y}$ vanishes. But then $\mathbf{y}$ cannot be an extremal vector. However, $\mathbf{Ay} = \lambda\mathbf{y}$ and therefore $\mathbf{y}$ is extremal unless $\mathbf{y} = 0$. Thus $\mathbf{x} = c\mathbf{z}$ and $\lambda$

*has only one linearly independent eigenvector whose components can all be taken positive.*

We now show that $\lambda$ is a simple root of $\Phi(z)$. It is enough to show that $\Phi'(\lambda) \neq 0$. But $\Phi'(\lambda)$ is the trace of the matrix $\mathbf{Q}$ that is adjoint to $\lambda\mathbf{I} - \mathbf{A}$. Since the rank of $\lambda\mathbf{I} - \mathbf{A}$ is $n - 1$, $\mathbf{Q}$ is not the null matrix. Further $(\lambda\mathbf{I} - \mathbf{A})\mathbf{Q} = 0$. This implies that every nonvanishing column of $\mathbf{Q}$ has elements of the same sign (since it is an eigenvector of $\mathbf{A}$ with eigenvalue $\lambda$). However, the relation $\mathbf{Q}(\lambda\mathbf{I} - \mathbf{A}) = 0$ implies the same of the rows of $\mathbf{Q}$. Thus all elements of $\mathbf{Q}$ have the same sign and hence $\Phi'(\lambda) = \text{trace} (\mathbf{Q}) \neq 0$. The proof of the theorem is complete.

Notice that (11) is an interesting maxmin property of the eigenvalue. Of course, *when* $\mathbf{A}$ *is a transition probability matrix* $\lambda = 1$. For

$$\min_i \frac{\sum_j a_{i,j}x_j}{x_i} \leq 1 \tag{15}$$

for any $\mathbf{x} \neq 0$, $\mathbf{x} \geq 0$, since $\sum_i a_{i,j} = 1$. However, 1 is attained by taking $x_1 = \cdots = x_n \neq 0$. *The left eigenvector* $\mathbf{x}$ *of* $\mathbf{A}$ *with eigenvalue one can be taken as the initial probability distribution of a Markov chain with transition matrix* $\mathbf{A}$ *if it is normed so that* $\sum_j x_i = 1$. In fact, this chain will then clearly be a stationary Markov chain.

The next result is an interesting comparison theorem.

**Theorem 2:** *Let* $\mathbf{A} = (a_{i,j})$ *be an irreducible matrix with non-negative elements and* $\mathbf{B} = (b_{i,j})$ *a matrix with complex elements,* $|b_{i,j}| \leq a_{i,j}, i, j = 1, \ldots, n$. *Let* $\lambda$ *be the maximal eigenvalue of* $\mathbf{A}$ *and* $\beta$ *an arbitrary eigenvalue of* $\mathbf{B}$. *Then* $|\beta| \leq \lambda$. *If equality holds* $\beta = \lambda e^{i\varphi}$ *and* $\mathbf{B}$ *can be written*

$$\mathbf{B} = e^{i\varphi}\mathbf{D}\mathbf{A}\mathbf{D}^{-1} \tag{16}$$

*where* $\mathbf{D}$ *is a diagonal matrix whose diagonal elements have absolute value one; further then* $|b_{i,j}| = a_{i,j}$.

Now

$$\beta\mathbf{x}' = \mathbf{B}\mathbf{x}' \tag{17}$$

implies

$$|\beta|\mathbf{x}^{*\prime} \leq \mathbf{B}^*\mathbf{x}^{*\prime} \leq \mathbf{A}\mathbf{x}^{*\prime} \tag{18}$$

so that $|\beta| \leq \lambda_{\mathbf{x}^*} \leq \lambda$. If $|\beta| = \lambda$, $\mathbf{x}^*$ is an extremal vector of $\mathbf{A}$ and

$$|\beta|\mathbf{x}^{*\prime} = \mathbf{A}\mathbf{x}^{*\prime}, \mathbf{x}^* > 0. \tag{19}$$

But then $\mathbf{B}^* = \mathbf{A}$, $|b_{i,j}| = a_{i,j}$. Now $\mathbf{x}' = \mathbf{D}\mathbf{x}^{*'}$ where $\mathbf{D}$ is a diagonal matrix whose diagonal entries have absolute value one. Set $\beta = \lambda e^{i\varphi}$. Then

$$|\beta|\mathbf{x}^{*'} = \mathbf{C}\mathbf{x}^{*'} \tag{20}$$

where

$$\mathbf{C} = e^{-i\varphi}\mathbf{D}^{-1}\mathbf{B}\mathbf{D}, \quad \mathbf{C}^* = \mathbf{B}^* = \mathbf{A}. \tag{21}$$

But $\mathbf{C}\mathbf{x}^{*'} = \mathbf{C}^*\mathbf{x}^{*'}$ and since $\mathbf{x}^* > 0$

$$\mathbf{C} = \mathbf{C}^*, \quad \mathbf{C} = \mathbf{A}, \quad \mathbf{B} = e^{i\varphi}\mathbf{D}\mathbf{A}\mathbf{D}^{-1}. \tag{22}$$

Theorem 2 can be used to obtain the following result.

**Theorem 3:** *Let $\mathbf{A}$ be an irreducible non-negative matrix. Suppose there are $k$ roots of $\Phi(z) = 0$ of maximal absolute value. Then they must all be simple and of the form $\lambda e^{2\pi i j/k}(j = 1, \ldots, k)$. The set of all $n$ roots of $\mathbf{A}$ are invariant under a rotation in the complex plane about zero through an angle of $2\pi/k$ but not through any smaller angle. By appropriate relabeling of rows (and correspondingly of columns) $\mathbf{A}$ can be put in the form*

$$\begin{bmatrix} 0 & \mathbf{A}_{12} & 0 & \cdots & 0 \\ 0 & 0 & \mathbf{A}_{23} & \cdots & 0 \\ 0 & 0 & 0 & \cdots & 0 \\ \cdot & \cdot & \cdot & \cdots & \cdot \\ 0 & 0 & 0 & \cdots & \mathbf{A}_{k-1,k} \\ \mathbf{A}_{k1} & 0 & 0 & \cdots & 0 \end{bmatrix}. \tag{23}$$

Suppose there are $k$ roots of maximal absolute value $\lambda$

$$\alpha_j = \lambda e^{i\varphi_i} \qquad 0 = \varphi_1 \le \varphi_2 \le \cdots \le \varphi_k < 2\pi. \tag{24}$$

The assumptions of Theorem 2 are satisfied with $\mathbf{B} = \mathbf{A}$ and $\beta = \alpha_j$ and thus there is a diagonal matrix $\mathbf{D}_j$ such that

$$\mathbf{A} = e^{i\varphi_i}\mathbf{D}_j\mathbf{A}\mathbf{D}_j^{-1}. \tag{25}$$

Since $\lambda$ is a simple root of $\Phi(z) = 0$ so are all the roots $\alpha_j\ j = 1, \ldots, k$. Thus the $\alpha_j$ are all distinct. Using (25) we see that

$$\mathbf{A} = e^{i(\varphi_i \pm \varphi_{i'})}\mathbf{T}\mathbf{A}\mathbf{T}^{-1}(\mathbf{T} = \mathbf{D}_j\mathbf{D}_{j'}^{\pm 1}) \tag{26}$$

so that the numbers $\lambda e^{i(\varphi_i \pm \varphi_{i'})}$ are eigenvalues of $\mathbf{A}$. But this implies that

$$\varphi_j = (j-1)\frac{2\pi}{k}. \tag{27}$$

The invariance of the set of eigenvalues through a rotation of $\frac{2\pi}{k}$ about the origin follows from (27).

Rearrange the rows and columns of the matrices so that $\mathbf{D}_2$ has the form

$$
\mathbf{D}_2 = \begin{bmatrix} \mathbf{I}_1 e^{i\delta_1} & & & \mathbf{0} \\ & \mathbf{I}_2 e^{i\delta_2} & & \\ & & \cdot & \\ & & & \cdot \\ \mathbf{0} & & & \mathbf{I}_g e^{i\delta_g} \end{bmatrix} \tag{28}
$$

where the $\mathbf{I}_i$'s are identity matrices and the $\delta_i$'s differ from each other mod $2\pi$. Carry out the corresponding decomposition of $\mathbf{A}$ into $g^2$ submatrices $\mathbf{A}_{ij}$. Equation (25) in the case $j = 2$ can then be rewritten

$$
\mathbf{A}_{ij} = e^{i\left(\frac{2\pi}{k} + \delta_i - \delta_j\right)} \mathbf{A}_{ij} \tag{29}
$$

so that $\mathbf{A}_{ij} = 0$ when $\frac{2\pi}{k} + \delta_i \neq \delta_j$ mod $2\pi$. Thus there is at most for each $i$ one $j$ for which $\mathbf{A}_{ij} \neq 0$. On the other hand there must be one since $\mathbf{A}$ is assumed to be irreducible. Thus given $\delta_i$ there is precisely one $j$ such that $\delta_j = \frac{2\pi}{k} + \delta_i$. If the rows (and columns) have been ordered properly

$$
\delta_j = \delta_1 + (j - 1)\frac{2\pi}{k} \quad j = 1, \ldots, k = g \tag{30}
$$

and $\mathbf{A}$ has the form (23).

We have already remarked that one is always an eigenvalue of maximal absolute value if $\mathbf{A}$ is a transition probability matrix. Let us see what the probabilistic meaning of other eigenvalues of absolute value one is when $\mathbf{A}$ is irreducible. If there are $k$ eigenvalues of absolute value one, Theorem 3 states that there are $k$ submatrices such that $\mathbf{A}$ can be laid out in block cyclic form (23). Call the distinct sets of row labelings (or sets of states) corresponding to $\mathbf{A}_{12}, \ldots, \mathbf{A}_{k1}$ the sets of states $S_1, \ldots, S_k$. Consider the natural order of the sets of states induced by the integer indexing with the convention that 1 follows $k$. From (23) it is clear that one can go from a state in set $S_i$ in one step with positive probability only to states in the following set of states. Consequently one can only return to $S_i$ after leaving $S_i$ with positive probability in a multiple of $k$ steps. It is natural to call such states periodic with period $k$. Theorem 3 tells us that if an irreducible transition probability matrix has $k$ eigenvalues of absolute value one, the states are all periodic with period $k$. We shall discuss periodic states again in section c.

A simple illustrative example is given by the circulant matrices, that is, matrices of the form

$$\mathbf{A} = \begin{bmatrix} a_0 & a_1 & \cdots & a_{n-1} \\ a_{n-1} & a_0 & \cdots & a_{n-2} \\ \cdot & & \cdot & \\ \cdot & & \cdot & \\ \cdot & & \cdot & \\ a_1 & a_2 & \cdots & a_0 \end{bmatrix} \tag{31}$$

Since matrices with non-negative elements are desired, the $a_j \geq 0$. However, most of our computations are valid without this restraint. Notice that the matrix $\mathbf{A}$ is a polynomial

$$\mathbf{A} = \sum_{k=0}^{n-1} a_k \mathbf{J}^k \tag{32}$$

in the matrix $\mathbf{J}$

$$\mathbf{J} = \begin{bmatrix} 0 & 1 & 0 & \cdots & 0 \\ 0 & 0 & 1 & \cdots & 0 \\ \cdot & & & \cdot & \\ \cdot & & & \cdot & \\ \cdot & & & \cdot & \\ 0 & 0 & 0 & \cdots & 1 \\ 1 & 0 & 0 & \cdots & 0 \end{bmatrix} \tag{33}$$

with $\mathbf{J}^0 = \mathbf{I}$, the identity matrix. We need therefore only obtain the eigenvalue and eigenvector structure of $\mathbf{J}$ in order to get that of $\mathbf{A}$. The eigenvalues of $\mathbf{J}$ are $\lambda_j = e^{2\pi ij/n}$ with corresponding right eigenvectors

$$x^{(j)} = \begin{bmatrix} 1 \\ e^{2\pi ij/n} \\ \cdot \\ \cdot \\ \cdot \\ e^{2\pi ij(n-1)/n} \end{bmatrix} \frac{1}{\sqrt{n}}, \qquad j = 0, 1, \ldots, n-1. \tag{34}$$

The eigenvalues of $\mathbf{A}$ are then $\lambda_j = a(e^{\frac{2\pi ij}{n}}) = \sum_{k=0}^{n-1} a_k e^{\frac{2\pi ikj}{n}}$. Notice that a maximal eigenvalue of $\mathbf{A}$ is $\lambda_0 = \sum_k a_k$ since the $a_k$'s are all non-negative. It will be simple as long as one $a_k$ other than $a_0$ is positive. The left eigenvector with maximal eigenvalue is $\dfrac{1}{\sqrt{n}}(1,1,\ldots,1)$.


# c. Limit Properties for Markov Chains

Let us now consider a Markov chain (finite or infinite state) with stationary transition mechanism. Let $\mathbf{P} = (p_{i,j}; i, j = 1, 2, \ldots)$, $p_{i,j} \geq 0$, $\sum_j p_{i,j} = 1$, be the transition probability matrix of the chain. Our primary interest is in the asymptotic behavior of the $n$-step transition probabilities $p_{i,j}^{(n)}$, $\mathbf{P}^n = (p_{i,j}^{(n)})$, as $n \to \infty$. An interesting and illuminating classification of the states of the chain will be introduced.

The state $k$ can be reached from the state $j$ if there is an integer $n \geq 1$ such that $p_{j,k}^{(n)} > 0$. We shall call *a class of states $C$ of the chain closed if $p_{j,k} = 0$ whenever $j$ is in $C$ and $k$ is outside.* Thus no state outside a closed class $C$ can be reached from any state in $C$. *A closed class of states is minimal if it contains no proper closed subset of states.* Notice that the statement that the matrix $\mathbf{P}$ is irreducible (see section b) is equivalent to the class of all states of the process being a minimal closed class of states. We might therefore call such a Markov chain an irreducible chain.

Consider now a fixed state $j$ of the chain and suppose the system studied is in state $j$ at time zero. Let $f_j^{(n)}$ then be the probability that the first return to $j$ occurs at time $n \geq 1$. Note that

$$f_j^{(1)} = p_{j,j}$$
$$f_j^{(2)} = p_{j,j}^{(2)} - f_j^{(1)}p_{j,j} \tag{1}$$

and generally

$$f_j^{(n)} = p_{j,j}^{(n)} - f_j^{(1)}p_{j,j}^{(n-1)} - \cdots - f_j^{(n-1)}p_{j,j}. \tag{2}$$

The sum

$$f_j = \sum_{n=1}^{\infty} f_j^{(n)} \tag{3}$$

is the probability that the system ever returns to the state $j$. If $f_j = 1$, the time for a first return to state $j$ is a well-defined random variable. The time required for a first return from state $j$ to state $j$ is called the *recurrence time for the state $j$.* If $f_j = 1$ the first moment

$$\mu_j = \sum_{n=0}^{\infty} nf_j^{(n)} \tag{4}$$

is the *mean recurrence time for state j*. Note that if $f_j < 1$, there is a positive probability the system will never return to $j$.

We shall introduce a classification of the states of the chain in terms of their recurrence time. The *state j is called a transient state if return to j is not sure, that is, $f_j < 1$. If return to j (from j) is sure ($f_j = 1$) the state is a recurrent state. A recurrent state with infinite mean recurrence time $\mu_j = \infty$ is called a null state. A state j is periodic with period t if return is impossible except in perhaps t, 2t, 3t, . . . steps and t is the greatest integer larger than one with this property. A recurrent state that is neither null nor periodic is called a persistent state.*

It should be noted that the study of Markov chains with an infinite number of states was initiated by A. Kolmogorov [45]. Further work was carried on by Doeblin [10], Doob [11], Chung [8], Feller [17], and others. The terminology used in this section is due to W. Feller [17].

The recurrence time distribution of a state of the chain can be studied by making use of generating functions. Equations (2) relate the probabilities $f_j^{(n)}$ of first return and the ordinary transition probabilities $p_{j,j}^{(n)}$. Suppose we let

$$f_j^{(0)} = 0, \; p_{j,j}^{(0)} = 1 \tag{5}$$

and introduce the generating functions

$$F(s) = \sum_{n=0}^{\infty} f_j^{(n)} s^n \tag{6}$$

$$G(s) = \sum_{n=0}^{\infty} p_{j,j}^{(n)} s^n. \tag{7}$$

Equations (1) and (2) can then be written in terms of the generating functions as

$$G(s) - 1 = F(s)G(s) \tag{8}$$

or

$$G(s) = \frac{1}{1 - F(s)}. \tag{9}$$

From this it is clear that $j$ is a *recurrent state if and only if*

$$\lim_{s \to 1-} G(s) = \sum_{n=0}^{\infty} p_{j,j}^{(n)} = \infty. \tag{10}$$

Consider as an example the unbounded random walk on the lattice points $i = 0, \pm 1, \pm 2, \ldots$ of the line with probability $p > 0$ of going one step to the right and $q = 1 - p > 0$ of going one step to the left. Suppose we start at a state (say $i = 0$) at time zero. The probability of returning in $n$ steps to zero is

$$p_{0,0}^{(n)} = \begin{cases} 0 \text{ if } n \text{ is odd} \\ \binom{n}{\frac{n}{2}} p^{\frac{n}{2}} q^{\frac{n}{2}} \text{ if } n \text{ is even.} \end{cases} \tag{11}$$

Thus it is already clear that the state (and hence every state) has period two. The generating function

$$
\begin{aligned}
G(s) &= \sum_{n=0}^{\infty} \binom{2n}{n} p^n q^n s^{2n} \\
&= \sum_{n=0}^{\infty} (-1)^n 2^{2n} \binom{-\frac{1}{2}}{n} p^n q^n s^{2n} \\
&= (1 - 4pqs^2)^{-\frac{1}{2}}.
\end{aligned}
\tag{12}
$$

Now

$$\lim_{s \to 1-} G(s) = (1 - 4pq)^{-\frac{1}{2}} \tag{13}$$

which is finite if $p \neq q$ and infinite if $p = q = \frac{1}{2}$. Thus if $p \neq q$, the states are all transient. This is not unexpected since if $p > q$ ($p < q$) we would expect a pronounced drift to the right (left). However, if there is balance, $p = q = \frac{1}{2}$, the states are recurrent. Now

$$F(s) = 1 - (1 - 4pqs^2)^{\frac{1}{2}} \tag{14}$$

so that when $p = q = \frac{1}{2}$, the mean recurrence time is given by

$$\lim_{s \to 1-} F'(s) = \lim_{s \to 1-} s(1 - s^2)^{-\frac{1}{2}} = \infty. \tag{15}$$

This is somewhat surprising for the random walk can be thought of in terms of a succession of independent tossings of fair coins. At time $n$ the location of the particle in the random walk is equal to the excess of heads over tails in $n$ tossings. Location zero corresponds to equilibrium, an equal number of heads and tails. Since the states are recurrent, we come back to equilibrium with probability one as expected. But the fact that the mean recurrence time is infinite implies that one may have to wait a very long time before one comes back to equilibrium.

Consider now an irreducible Markov chain. Let $j,k$ be any two states of the chain. Since the chain is irreducible one can go from $j$ to $k$ and from $k$ to $j$ with positive probability. For suppose $k$ is inaccessible from $j$. Then by deleting $k$ and all states that lead into $k$ with positive probability we would obtain a proper closed subchain contrary to the assumption that the original chain is irreducible. Let $N$ be the length of the shortest path from $j$ to $k$ with positive probability. Similarly let $M$ be the length of the shortest path from $k$ to $j$ with positive probability. Set $p_{j,k}^{(N)} = \alpha > 0$, $p_{k,j}^{(M)} = \beta > 0$. Then for any positive $n$ it is clear that

$$
\begin{aligned}
p_{j,j}^{(n+N+M)} &\geq p_{j,k}^{(N)} p_{k,k}^{(n)} p_{k,j}^{(M)} = \alpha\beta p_{k,k}^{(n)} \\
p_{k,k}^{(n+N+M)} &\geq p_{k,j}^{(M)} p_{j,j}^{(n)} p_{j,k}^{(N)} = \alpha\beta p_{j,j}^{(n)}.
\end{aligned}
\tag{16}
$$

Thus the sequences $p_{j,j}^{(n)}$, $p_{k,k}^{(n)}$ have the same asymptotic behavior. We already know that $j$ is recurrent if and only if $\sum_{n=0}^{\infty} p_{j,j}^{(n)} = \infty$. But the inequalities (16) imply the equivalence of the divergence of $\Sigma p_{j,j}^{(n)}$ and $\Sigma p_{k,k}^{(n)}$. Thus *the states of an irreducible chain are either all recurrent or all transient.* Now suppose $j$ is periodic with period $t$. The number $N + M$ is divisible by $t$ since a return to $j$ is possible in $N + M$ steps. The inequalities (16) then imply that $k$ is periodic with period $t$. Therefore, *if one state in an irreducible chain is periodic with period $t$, all the states are periodic with period $t$.* A theorem of Erdös, Feller, and Pollard indicates that

$$
p_{j,j}^{(n)} \to \frac{1}{\mu_j} = u_j
\tag{17}
$$

as $n \to \infty$ if $j$ is a recurrent nonperiodic state. Here, of course, $\mu_j$ is the mean recurrence time of the state $j$. *If $j$ is a null state $u_j = 0$ and if $j$ is a persistent state then $u_j > 0$.* Let us assume this result for the moment. The Erdös, Feller, Pollard theorem will be derived later on. The Erdös, Feller, Pollard theorem and inequalities (16) imply that *all the states of an irreducible chain are persistent or none of them are.*

Let us briefly discuss the case of an irreducible chain with states of period $t$. Let $j$ and $k$ be any two states of the chain. Since the chain is irreducible there are integers $N$, $M$ such that $p_{j,k}^{(N)} > 0$, $p_{k,j}^{(M)} > 0$. Since $p_{j,j}^{(N+M)} \geq p_{j,k}^{(N)} p_{k,j}^{(M)}$, the integer $N + M$ is divisible by $t$. Hence if $N$, $N'$ are the lengths of two paths of positive probability from $j$ to $k$, they must have the same remainder after division by $t$. For fixed $j$ each state $k$ corresponds to a remainder $\nu$, $0 \leq \nu \leq t - 1$, such that a transition from $j$ to $k$ with positive probability is possible only in $\nu, \nu + t, \nu + 2t, \ldots$ steps. Take $j = 1$ and classify all the states into disjoint sets $S_0, S_1, \ldots, S_{t-1}$ where $k$ belongs to $S_\nu$ if $p_{1,k}^{(N)} > 0$ implies that $N = \nu + nt$. Consider the sets $S_\nu$ in their natural indexed order with the convention that $S_0$ follows $S_{t-1}$. It is then clear that a one-step transition from a state in $S_\nu$ will lead to a state in the following set of states. Notice that a $t$-step transition leads back to a state in the same set. If we therefore consider the matrix of transition probabilities $\mathbf{P}^t$, each set of states $S_\nu$ is a closed irreducible nonperiodic class of states.

Consider an irreducible nonperiodic Markov chain. Let us now examine the asymptotic behavior of the transition probabilities $p_{j,k}^{(n)}$ as $n \to \infty$ where $j$ and $k$ may be distinct states. We already know that the irreducible character of the chain implies that all states must be of the same type. Now

$$p_{j,k}^{(n)} = \sum_{m=1}^{n} f_{j,k}^{(m)} p_{k,k}^{(n-m)} \tag{18}$$

where $f_{j,k}^{(m)}$ is the probability of a first passage from $j$ to $k$ in precisely $m$ steps. *If the states of the chain are transient* $p_{k,k}^{(n)} \to 0$ *as* $n \to \infty$ *for all* $k$ and hence by relation (18) $p_{j,k}^{(n)} \to 0$ *as* $n \to \infty$ *for all* $j,k$ ( $\sum\limits_{m=1}^{\infty} f_{j,k}^{(m)} \leq 1$ ).

*If the states of the chain are null states,* $p_{k,k}^{(n)} \to \dfrac{1}{\mu_k} = \dfrac{1}{\infty} = 0$ as $n \to \infty$ and the same argument indicates that $p_{j,k}^{(n)} \to 0$ *as* $n \to \infty$. Now consider the case of persistent states. Since the states are recurrent $\sum\limits_{m} f_{j,k}^{(m)} = 1$ for otherwise return to $j$ from $j$ would not occur with probability one. Further by (17) $p_{j,j}^{(n)} \to \dfrac{1}{\mu_j} = u_j > 0$ as $n \to \infty$ and thus relation (18) implies $p_{j,k}^{(n)} \to u_k > 0$ as $n \to \infty$. We shall now show that *if the chain is irreducible and nonperiodic, there is an invariant instantaneous distribution if and only if the states are persistent, in which case the distribution is unique and given by* $\{u_k\}$. Let $\{v_k\}$ be an instantaneous invariant distribution, that is, a probability distribution satisfying

$$v_j = \sum_i v_i p_{i,j}. \tag{19}$$

Relation (19) implies that

$$v_j = \sum_i v_i p_{i,j}^{(n)} \tag{20}$$

for all $n$. On letting $n \to \infty$ we have

$$v_j = \sum_i v_i u_j = u_j \tag{21}$$

so that if there is an invariant distribution it is uniquely given by $\{u_j\}$. The states of the process must therefore be persistent. All that is now required is to show that $\{u_j\}$ is in fact a probability distribution satisfying (19) if the states are persistent. Now $\sum_k u_k \leq 1$ since $\sum_k p_{j,k}^{(n)} = 1$ and $u_k = \lim_{n \to \infty} p_{j,k}^{(n)}$. Now $p_{j,k}^{(n+1)} = \sum_\nu p_{j,\nu}^{(n)} p_{\nu,k}$. On letting $n \to \infty$ the inequality

$$u_k \geq \sum_\nu u_\nu p_{\nu,k} \tag{22}$$

is obtained. On summing over $k$ the same quantity $\sum_\nu u_\nu$ is obtained and hence the inequalities must have been equalities. But then $\{u_\nu\}$ is a nontrivial solution of (19) since the $u_k > 0$. Setting $v_k = u_k(\Sigma u_\nu)^{-1}$ an invariant distribution is obtained. But we have already seen that an invariant distribution if it exists must be $\{u_k\}$. Thus $\sum_\nu u_\nu = 1$.

Notice that existence of an invariant distribution means that one is a left eigenvalue of $\mathbf{P}$ with corresponding eigenvector a probability vector. Periodic states imply the existence of other eigenvalues of $\mathbf{P}$ of absolute value one.

The Erdös, Pollard, Feller, theorem relates to the asymptotic behavior of a sequence $p_{j,j}^{(n)}$ ($n \to \infty$) satisfying relation (2). We now state and prove this theorem. In the following proof think of $f_n$ as the probability of first return to a state in $n$ steps and of $u_n$ as the transition probability of return (not necessarily a first return) to the same state in $n$ steps.

**Theorem:** *Let* $\{f_n\}$, $n = 0, 1, \ldots$, *be such that* $f_0 = 0$, $f_n \geq 0$, $\Sigma f_n = 1$ *and the greatest common divisor of the integers $n$ for which $f_n > 0$ is one. Set* $u_0 = 1$ *and for* $n \geq 1$ *let*

$$u_n = f_1 u_{n-1} + f_2 u_{n-2} + \cdots + f_n u_0. \tag{23}$$

*Then* $u_n \to 1/\mu$ *where* $\mu = \Sigma n f_n$ $(u_n \to 0$ *if* $\Sigma n f_n = \infty)$ *as* $n \to \infty$.

Let $r_n = f_{n+1} + f_{n+2} + \cdots$. Then $r_0 = 1$ and $f_j = r_{j-1} - r_j$, $j = 1, 2, \ldots$. Using this relation between the $f$'s and $r$'s, equation (23) can be rewritten as

$$A_n = \sum_{j=0}^{n} r_j u_{n-j} = \sum_{j=0}^{n-1} r_j u_{n-1-j} = A_{n-1} = \cdots = r_0 u_0 = A_0 = 1. \tag{24}$$

Now $u_0 = 1$ and hence by equation (23) $0 \leq u_n \leq 1$ for all $n$. Let $\lambda = \lim_{n \to \infty} \sup u_n$. Further let $j > 0$ be such that $f_j > 0$. We show that if $n_\nu$ is an increasing sequence such that $u_{n_\nu} \to \lambda$, then $u_{n_\nu - j} \to \lambda$. For if this were not so there would be arbitrarily large $n$ in the sequence $n_\nu$ with $u_{n-j} < \lambda' < \lambda$. Take $N$ so large that $r_N = \sum\limits_{N+1}^{\infty} f_j < \varepsilon$, $\varepsilon > 0$. Since $u_k \leq 1$, for $n > N$

$$u_n \leq f_0 u_n + f_1 u_{n-1} + \cdots + f_N u_{n-N} + \varepsilon. \tag{25}$$

For sufficiently large $n$ of this sort we would then have

$$\lambda - \varepsilon < u_n < (f_0 + \cdots + f_{j-1} + f_{j+1} + \cdots + f_N)(\lambda + \varepsilon) \\ + f_j \lambda' + \varepsilon \tag{26}$$
$$\leq (1 - f_j)(\lambda + \varepsilon) + f_j \lambda' + \varepsilon < \lambda + 2\varepsilon - f_j(\lambda - \lambda')$$

and this leads to a contradiction if $f_j(\lambda - \lambda') > 3\varepsilon$. Therefore $u_{n_\nu - j} \to \lambda$. A repetition of this argument indicates that $u_{n_\nu - 2j} \to \lambda$ and generally $u_{n_\nu - xj} \to \lambda$ for any fixed positive integer $x$. Similarly if $\gamma = \lim\inf\limits_{n \to \infty} u_n$ and $n_\nu$ is any sequence for which $u_{n_\nu} \to \gamma$, then $u_{n_\nu - xj} \to \gamma$ for any positive integer $x$. Now consider all the integers $j$ for which $f_j > 0$. There is a finite subcollection $a$, $b$, $c$, $\ldots$, $m$ of these integers with greatest common divisor 1. Now by an extension of the argument given above if $u_{n_\nu} \to \lambda$, then $u_{n_\nu - xa - yb - \ldots - wm} \to \lambda$ for all fixed integers $x$, $y$, $\ldots$, $w > 0$. But every integer $k$ greater than the product $abc \cdots m$ can be represented in the form $k = xa + yb + \cdots + wm$ with $x$, $y$, $\ldots$, $w > 0$. Thus $u_{n_\nu - k} \to \lambda$ for every $k > ab \cdots m$. The analogous result for convergence to $\gamma$ holds if $u_{n_\nu} \to \gamma$. Now let $u_{n_\nu} \to \lambda$. Then $u_{n_\nu - k} \to \lambda$ for $k > ab \cdots m$ and taking $n = n_\nu - ab \cdots m - 1$ we have

$$1 \geq r_0 u_n + r_1 u_{n-1} + \cdots + r_N u_{n-N}. \tag{27}$$

On letting $n_\nu \to \infty$ the inequality $1 \geq \lambda(r_0 + \cdots + r_N)$ is obtained. Since this is true for all $N$, $1 \geq \lambda\mu$ since $\mu = \Sigma r_j$. Notice that this implies that $\lambda = \lim u_n = 0$ if $\mu$ is infinite. Suppose $\mu$ finite. Take $N$ sufficiently large so that $\sum\limits_{j > N} r_j < \varepsilon$, $\varepsilon > 0$. Let $n_\nu$ be an increasing sequence such that $u_{n_\nu} \to \gamma$. Take $n = n_\nu - ab \cdots m - 1$. Then

$$1 \leq r_0 u_n + \cdots + r_N u_{n-N} + \varepsilon \tag{28}$$

and on letting $n_\nu \to \infty$ we obtain $1 \leq (r_0 + \cdots + r_N)\gamma + \varepsilon$ and hence $\mu\gamma \geq 1$. Since it was already shown that $\mu\lambda \leq 1$, the equality $\lambda = \gamma = \dfrac{1}{\mu}$ must hold.

Relation (17) is obtained by applying the above theorem to the equations (2). For the asymptotic behavior of transition probability of periodic states see problem III.14.

## d. Functions of a Markov Chain

The structure of a Markov chain $X(m)$ has been discussed.* However, in various contexts, it may be that the experimenter does not observe the process $X(m)$ but rather a derived process $Y(m) = f(X(m))$ where $f$ is a given function on the state space of the Markov chain $X(m)$. The states $i$ of the original process on which $f$ equals some fixed constant are collapsed into a single state of the new process $Y(m)$. We shall call the collection of states on which $f$ takes the value $\alpha$ the *set of states* $S_\alpha$. Of course, only nonempty sets of states are of interest. One would like to know whether the new process is Markovian. The following simple example indicates that this is generally not the case. Let the initial process $X(m)$ be a Markov chain with the three states 1, 2, 3, transition probability matrix

$$\mathbf{P} = \begin{bmatrix} 0 & \frac{1}{2} & \frac{1}{2} \\ \frac{1}{3} & \frac{1}{4} & \frac{5}{12} \\ \frac{2}{3} & \frac{1}{4} & \frac{1}{12} \end{bmatrix} \tag{1}$$

and initial probability vector

$$\mathbf{p} = (\tfrac{1}{3}, \tfrac{1}{3}, \tfrac{1}{3}). \tag{2}$$

The process $X(m)$ is stationary since $\mathbf{p}$ is a left eigenvector of $\mathbf{P}$ with eigenvalue one. Collapse the set of states $S$ consisting of 1, 2 into one state. Then

$$P(X(m+2)\epsilon\, S,\ X(m+1)\epsilon\, S | X(m)\epsilon\, S) = {}^{29}\!\!/_{96}$$
$$\neq P(X(n+1)\epsilon\, S | X(n)\epsilon\, S)^2 = (\tfrac{13}{24})^2 \tag{3}$$

so that the new process is not Markovian. It is natural to look for conditions on the function $f$ and the probability structure of the Markov chain $X(m)$ that will imply that the new process $Y(m) = f(X(m))$ is Markovian. For if $Y(m)$ is Markovian, we can apply the powerful techniques that are relevant in the analysis of such processes to the computation

---

* Every so often we shall write $X(m)$ instead of $X_m$. There will however be no confusion as it will always be clear what is meant from the context.

of aspects of the probability structure of $Y(m)$. In section a we have already noted that the transition probabilities of a Markov chain satisfy the Chapman-Kolmogorov equation. Such a condition might therefore possibly be provided by insisting that the transition probabilities of the $Y(m)$ process satisfy the Chapman-Kolmogorov equation. However, an example of P. Lévy [52] indicates that this condition does not generally imply that the new process $Y(m)$ is Markovian. We now give an example that is a simplification of P. Lévy's to illustrate this.

Let $X(m) = (Y(m), Y(m - 1))$, $m = 1, 2, \ldots$, be a Markov chain with transition probabilities

$$P(Y(m + 2) = u_2 | Y(m + 1) = u_1, Y(m) = u_0)$$
$$= \frac{1}{r} \left\{ 1 - \cos \left[ \frac{2\pi}{r} (2u_2 - u_1 - u_0) \right] \right\}$$
$$u_0, u_1, u_2 = 0, 1, \ldots, r - 1. \quad (4)$$

Thus $X(m)$ is a Markov chain with a state space consisting of the $r^2$ points $(i,j)$, $i, j = 0, \ldots, r - 1$. Let the initial probability distribution be

$$p_{u_0, u_1} = P[Y(0) = u_0, Y(1) = u_1] = \frac{1}{r^2}. \quad (5)$$

With this initial distribution $X(m)$ is stationary and persistent. The process $Y(m)$ is a function of $X(m)$ but is not Markovian (it might be appropriate to call it two-step Markovian) since the transition structure (4) depends on the two previous locations of the process. However, a direct computation indicates that the one-step transition probabilities of $Y(m)$ are of the form

$$P(Y(t) = u_t | Y(s) = u_s) = \frac{1}{r}, \quad 1 \leq s < t, \quad (6)$$

and they clearly satisfy the Chapman-Kolmogorov equation. We thus have a simple example of a non-Markovian process $Y(m)$ whose one-step conditional probabilities (6) satisfy the Chapman-Kolmogorov equation.

Let $X(m)$ be a Markov chain with transition probability matrix $\mathbf{P} = (p_{i,j})$. Take $f$ a given function on the integers. The requirement that we will impose on $f$ and $\mathbf{P}$ will be somewhat stronger than that spoken of above. It will be that $Y(m) = f(X(m))$ be Markovian whatever the initial probability distribution of $X(m)$. Let the sets of states of such a process $X(m)$ that are collapsed into single states $\alpha$ of $Y(m)$

by the function $f$ be denoted by $S_\alpha$. Take any initial distribution $w = (w_i)$, $w_i > 0$, for $X(m)$. If the new process $Y(m)$ is Markovian, then

$$P[Y(1) = \alpha, Y(2) = \beta, Y(3) = \gamma]$$
$$= \sum_{i\varepsilon S_\alpha, j\varepsilon S_\beta} w_i p_{i,j} p_{j,S_\gamma}$$
$$= P[Y(1) = \alpha]P[Y(2) = \beta|Y(1) = \alpha]P[Y(3) = \gamma|Y(2) = \beta]$$
$$= \sum_{i\varepsilon S_\alpha} w_i p_{i,S_\beta} \frac{\sum\limits_{u}\sum\limits_{v\varepsilon S_\beta} w_u p_{u,v} p_{v,S_\gamma}}{\sum\limits_{u} w_u p_{u,S_\beta}} \tag{7}$$

where

$$p_{i,S_\alpha} = \sum_{j\varepsilon S_\alpha} p_{i,j}. \tag{8}$$

The expression on the right of (7) is well defined as long as $p_{i,S_\beta} > 0$ for some $i$. We shall *call* a set of states $S_\alpha$ a *single entry set if* $p_{i,S_\alpha} > 0$ *for $i$ from at most one set of states $S_\beta$.* Single entry sets complicate the discussion and therefore *it will be assumed that none of the sets of states $S_\alpha$ are single entry sets.* A discussion of the problem allowing single entry sets in the case of Markov processes with a general state space is to be found in [68]. Consider any two states $i$, $i'$ in distinct sets of states $S_\alpha$, $S_{\alpha'}$ for which $p_{i,S_\beta}$, $p_{i',S_\beta} > 0$. Let all the $w_u$ with $u \neq i$, $i'$ tend to zero. Relation (7) then implies that

$$w_i \sum_{j\varepsilon S_\beta} p_{i,j} p_{j,S_\gamma} = w_i p_{i,S_\beta} \frac{\sum\limits_{v\varepsilon S_\beta}(w_i p_{i,v} + w_{i'} p_{i',v}) p_{v,S_\gamma}}{w_i p_{i,S_\beta} + w_{i'} p_{i',S_\beta}}. \tag{9}$$

But this is valid for all $w_i$, $w_{i'}$ only if

$$p_{i,S_\beta} \sum_{v\varepsilon S_\beta} p_{i',v} p_{v,S_\gamma} = p_{i',S_\beta} \sum_{v\varepsilon S_\beta} p_{i,v} p_{v,S_\gamma}. \tag{10}$$

Notice that (10) is obviously satisfied if $p_{i,S_\beta} = 0$. Equation (10) therefore holds for all $i$, $i'$, $\beta$, $\gamma$.

We shall show that *if all the collapsed sets of states $S_\alpha$ are not single entry sets, relation (10) is a necessary and sufficient condition for the new process $Y(m) = f(X(m))$ to be Markovian.* Since the necessity has already been proven, it is only necessary to prove it sufficient. Let $C_{S_\beta, S_\gamma}$ be the common value of

$$\sum_{v\varepsilon S_\beta} p_{i,v} p_{v,S_\gamma}/p_{i,S_\beta} \tag{11}$$

for all $i$ for which $p_{i,S_\beta} > 0$. Now

$$
\begin{aligned}
P[Y(1) &= \alpha_1, \ldots, Y(n) = \alpha_n] \\
&= \sum_{i_k \varepsilon S_{\alpha_k}} w_{i_1} p_{i_1,i_2} \cdots p_{i_{n-2},i_{n-1}} p_{i_{n-1},S_{\alpha_n}} \\
&= \sum_{i \varepsilon S_{\alpha_1}} w_i \frac{\sum\limits_{i \varepsilon S_{\alpha_1}} w_i p_{i,S_{\alpha_2}}}{\sum\limits_{i \varepsilon S_{\alpha_1}} w_i} C_{S_{\alpha_2},S_{\alpha_3}} \cdots C_{S_{\alpha_{n-1}},S_{\alpha_n}} \\
&= P[Y(1) = \alpha_1] P[Y(2) = \alpha_2 | Y(1) = \alpha_1] \cdots \\
&\qquad\qquad\qquad\qquad P[Y(n) = \alpha_n | Y(n-1) = \alpha_{n-1}] \quad (12)
\end{aligned}
$$

so that $Y(m)$ is Markovian. Notice that the transition mechanism from time 2 on is stationary since

$$
C_{S_\alpha,S_\beta} = P[Y(m+1) = \beta | Y(m) = \alpha] \tag{13}
$$

for $m \geq 2$. If the transition mechanism is to be stationary for all time, that is, for time 1 also,

$$
\sum_{i \varepsilon S_{\alpha_1}} w_i p_{i,S_{\alpha_2}} \Big/ \sum_{i \varepsilon S_{\alpha_1}} w_i = C_{S_{\alpha_1},S_{\alpha_2}} \tag{14}
$$

must hold for all $w_i > 0$. This leads to a condition somewhat more stringent than (10), namely for all $\alpha$, $\beta$

$$
p_{i,S_\beta} = C_{S_\alpha,S_\beta} \tag{15}
$$

when $i \varepsilon S_\alpha$. Condition (15) had already arisen in some work of B. Rankin [61] and a number of aggregation problems as they arise in econometrics [65]. One can readily see that condition (15) is more restrictive than (10). For if given any $\beta$

$$
p_{i,j} = q_{i,S_\beta} q_j \tag{16}
$$

for all $i$ and $j \varepsilon S_\beta$ relation (10) is satisfied while (15) need not be.

The corresponding problem in the case of a stationary Markov chain, that is, whether a function of a stationary Markov chain is Markovian, does not appear to have a simple answer generally. However, one can give a complete answer for certain special types of chains. Notice that we do not ask that the Markovian property be preserved for any initial distribution, only for an invariant initial probability vector. If the chain is irreducible there can be at most one such vector. Let $\mathbf{P}$ be the transition probability matrix of a chain and $\mathbf{p}$ a left invariant probability vector of $\mathbf{P}$. Assume that all the components of $\mathbf{p}$ are positive. Let $\mathbf{D}$ be the diagonal matrix with its $i$th diagonal entry $p_i$.

*A chain with transition matrix* **P** *and stationary instantaneous probability distribution* **p** *is said to be reversible if*

$$DP = P'D. \tag{17}$$

This is an appropriate name since relation (17) implies that the backward and forward transition probabilities are the same

$$P(X(m + 1) = j|X(m) = i) = P(X(m) = j|X(m + 1) = i). \tag{18}$$

For such chains evolution forwards and backwards in time are statistically indistinguishable. We shall show that *for reversible chains relation* (15) *is a necessary and sufficient condition for a function* $Y(m) = f(X(m))$ *of the chain to retain the Markovian property.* The sufficiency is already apparent. A demonstration of the necessity of the condition is only required. For convenience, the computation is carried out in the case of a chain $X(m)$ with a finite number of states. Let $u$ be the number of states of $X(m)$ and $v < u$ the number of distinct sets of states that $f$ collapses the $X$ state space into. We introduce $v \times u$ and $u \times v$ matrices **A**, **B** as follows. The elements of **B** are of the form

$$b_{i,j} = \begin{cases} 1 & \text{if } i \epsilon S_j \\ 0 & \text{otherwise} \end{cases} \tag{19}$$

while

$$A = (B'DB)^{-1}B'D. \tag{20}$$

The $n$-step transition probability matrix of the $Y(m)$ process is of the form

$$Q^{(n)} = AP^nB = (q_{i,j}^{(n)})$$
$$q_{i,j}^{(n)} = P(Y(t + n) = S_j|Y(t) = S_i). \tag{21}$$

If $Y(m)$ is Markovian, the Chapman-Kolmogorov equations are satisfied by the $Q^{(n)}$ and in particular we must have

$$Q^{(2)} = [Q^{(1)}]^2 \tag{22}$$

or

$$AP(I - BA)PB = 0. \tag{23}$$

But this implies that

$$B'DP(I - BA)PB = 0. \tag{24}$$

Because of the reversibility of the $X(m)$ process this last equation can be written as

$$B'P'D(I - BA)PB = 0. \tag{25}$$

The matrix $\mathbf{D}(\mathbf{I} - \mathbf{BA})$ is positive definite so that

$$\mathbf{D}(\mathbf{I} - \mathbf{BA}) = \mathbf{R}'\mathbf{R} \tag{26}$$

for some $u \times u$ matrix $\mathbf{R}$. Thus $(\mathbf{RPB})'(\mathbf{RPB}) = 0$ and hence $\mathbf{RPB} = 0$. But then

$$\mathbf{R}'\mathbf{RPB} = \mathbf{D}(\mathbf{I} - \mathbf{BA})\mathbf{PB} = 0 \tag{27}$$

and hence

$$(\mathbf{I} - \mathbf{BA})\mathbf{PB} = 0. \tag{28}$$

This last equation is simply relation (15) written in matrix notation.

## e. Problems

1. Show that if $X_n$ $n = 0, 1, 2, \ldots$ is a Markov chain, then

$$P[X_{n_1} = i_1 | X_{n_2} = i_2, \ldots, X_{n_k} = i_k] = P[X_{n_1} = i_1 | X_{n_2} = i_2]$$

   if $n_1 > n_2 > \cdots > n_k$. This is equivalent to the Markov property for a chain as stated in this chapter.

2. Let $\mathbf{P}$ be a transition probability matrix with column sums equal to one. Let $\mathbf{x}$ be a probability vector and $\mathbf{y}$ the probability vector generated from $\mathbf{x}$ by $\mathbf{y} = \mathbf{xP}$. Show that the entropy of $\mathbf{y}$ is greater than or equal to that of $\mathbf{x}$. What happens if the condition that column sums equal one is relaxed?

3. Show that there is no invariant probability vector for

$$p_{i,j} = \begin{cases} \lambda^{j-i}e^{-\lambda}/(j - i)! & \text{if } j \geq i \\ 0 & \text{otherwise.} \end{cases}$$

4. Find the eigenvalues of the transition probability matrix

$$\begin{pmatrix} 0 & 0 & \tfrac{1}{2} & \tfrac{1}{2} \\ 0 & 0 & \tfrac{1}{2} & \tfrac{1}{2} \\ \tfrac{1}{4} & \tfrac{3}{4} & 0 & 0 \\ \tfrac{3}{4} & \tfrac{1}{4} & 0 & 0 \end{pmatrix}.$$

5. In the case of a branching process $\{X_n\}$ show that $\text{Var}(X_{n+1}) = \mu\text{Var}(X_n) + \mu^{2n}\sigma^2$ where $\mu$ is the mean and $\sigma^2$ the variance of the $q_i$ distribution.

6. Let $X_1, X_2, \ldots$ be independent random variables with common distribution $P(X = j) = f_j$, $j = 0, 1, 2, \ldots$. Let $N$ be a non-negative random variable independent of the $X$'s with distribution

$P(N = m) = g_m$, $m = 0, 1, 2, \ldots$. Show that the distribution of the sum $X_1 + X_2 + \cdots + X_N = S_N$ (the sum of a random number of random variables) is given by $P(S_N = k) = \sum\limits_{m=0}^{\infty} g_m f_k^{(m*)}$ where $f_k^{(m*)}$ is the $k$-th element of the $m$-th convolution of the $f_k$ sequence with itself. Find the generating function of the probability distribution of $S_N$.

7. Consider an insurance company. Suppose that the number of deaths of people insured by the company in time $t$ is given by a Poisson distributed random variable with mean $\lambda t$. Further, let the amount of insurance to be paid up by the company on the death of an insured person be given by $P(X = j) = f_j$ in terms of dollars $j$. Assuming independence, find the generating function of the distribution of $S_{N(t)}$, the total amount of insurance paid out by the company in time $t$. The probability distribution of $S_{N(t)}$ is called a compound Poisson distribution.

8. Show that for any positive integer $n$ there is a probability distribution $\{g_k\}$ such that the sum of $n$ independent, identically distributed random variables with this distribution has the distribution of $S_{N(t)}$ referred to in the previous example. For this reason, a compound Poisson distribution is called infinitely divisible. See the book of Kolmogorov and Gnedenko [23] for a detailed discussion of infinitely divisible laws.

9. Consider a random walk with reflecting barriers at 0 and $a$. Let $p_j^{(n)} = P(X_n = j)$, $j = 0, 1, \ldots, a$. Examine the behavior of $p_j^{(n)}$ as $n \to \infty$. Does this behavior depend on the initial distribution? We say 0 and $a$ are reflecting barriers if $p_{0,1} = p_{a,a-1} = 1$.

10. Consider the preceding problem with an absorbing barrier at 0 and a reflecting barrier at $a$.

11. Consider a finite state Markov chain whose transition matrix has only positive elements. Show that all the states are persistent.

12. What is the left invariant probability vector of $\mathbf{P}$ if $\mathbf{P}$ is an irreducible transition probability matrix with column sums equal to one?

13. Let the first column of the transition probability matrix $\mathbf{P}$ of a Markov chain be $\{q_0, q_1, \ldots\}$ with $p_{i,i+1} = 1 - q_i$ for $i = 0$, $1, \ldots$. When is the chain irreducible? Show that if the chain is irreducible, the states are transient if and only if $\Sigma q_j < \infty$. Find out when the states are null and when they are persistent.

14. Show that if $j$ is a recurrent periodic state of period $t > 1$, then $p_{j,j}^{(m)} = 0$ if $m$ is not a multiple of $t$ and

$$p_{j,j}^{(st)} \to \frac{t}{\mu_j}$$

as $s \to \infty$ where $\mu_j$ is the mean recurrence time of state $j$. Hint: Apply the Erdös, Feller, Pollard theorem to $f_j^{(st)}$ and $p_{j,j}^{(st)}$.

15. Suppose **P** is an irreducible transition probability matrix. Show that $\lim\limits_{n \to \infty} \dfrac{1}{n} \sum\limits_{m=0}^{n-1} p_{j,j}^{(m)}$ exists and is equal to 0 if the states are transient and $1/\mu_j$ if the states are recurrent.

16. Let **P** be a transition probability matrix (not necessarily irreducible). Show that $\lim\limits_{n \to \infty} \dfrac{1}{n} \sum\limits_{m=0}^{n-1} p_{i,j}^{(m)}$ exists for all $i, j$. Evaluate the limit.

# Notes

1. Instead of using (a.4) to define Markov chains, property (a.9) or that given in example 1 could have been employed. In fact, these other approaches are more usual and might be considered more natural. In spite of this, property (a.4) appeared to be more convenient to use at this point. Specific types of Markov chains of interest are discussed in a variety of books on chains. We mention the book of Frechét [19] which discusses some of the early literature.

2. Section b is given because it was felt that an analysis of transition probability matrices from the point of view of their eigenvalue structure was proper as a valuable complement to the purely probabilistic approach. This is an analysis of matrices with non-negative elements of finite order and was originally developed in the work of Perron and Frobenius (see Gantmacher, vol. 2 [21]). Row sums of the matrices are not assumed to be equal because this condition is not required for the characterization of the eigenvalue structure. Matrices with non-negative elements have recently attracted interest in the analysis of flow networks. The element $a_{i,j}$ can be interpreted as the mass flowing per unit time from location $i$ to location $j$. Row sums equal would correspond to conservation of mass. Row sums unequal could be accounted for by creation or destruction of mass. See [5] for an interesting application to reactor problems. The paper of D. Rosenblatt [64] in part discusses the application of such matrix representations in the analysis of economic "input-output" problems.

3. The basic results on Markov chains presented in section c appear to be due to Kolmogorov [45] and Doeblin [10]. The presentation is essentially that given by Feller [17]. It should be noted that the terminology for states of a chain varies from one presentation to another. The terminology in Feller's second edition differs from that given in his first edition. Chung's terminology [8] differs from that given

in Feller. We have used the terminology given in Feller's first edition. The state classification given in his second edition is a little unpleasant since periodic states of an ergodic chain (see section b of Chapter V) are not called ergodic states. A much more extensive treatment of infinite state Markov chains and a discussion of the historical background can be found in Chung's monograph [8].

4. The question of necessary and sufficient conditions for a function of a general finite state Markov chain to be Markovian is still open. Further results on related problems can be found in [7] and [68]. Related aggregation problems are considered in the paper of D. Rosenblatt [65]. Recent work on conditions for a function of a general state Markov process with stationary transition mechanism to be Markovian can be found in the book of M. Rosenblatt [A14].

# IV

# PROBABILITY SPACES WITH AN INFINITE NUMBER OF SAMPLE POINTS

## a. Discussion of Basic Concepts

Thus far, we have concerned ourselves with probability spaces that contain a finite or at most a denumerable number of sample points. However, the natural probability spaces in many contexts will contain a nondenumerable number of sample points. The simplest situation of this sort arises when the result of an experimental measurement may be any real number, for the set of real numbers is nondenumerable (see [1]). Much of this chapter will be concerned with the introduction of concepts and tools relevant in such a general context. This presentation is motivated to a great extent by the discussion in the denumerable case. Sigma-fields and probability set functions will be introduced as they were in that simpler domain. Generally our object will be the understanding of these concepts and tools rather than a concern with subtleties and proofs. A detailed derivation of the results we present and discuss can be found in many books on measure theory (see [29], [53]). Typically measure theory is used as a basic tool in setting up the foundations of probability theory.

Consider an experiment. As before, there is a space $\Omega$ of points ("sample points") $w$, where the points are to be regarded as the elementary outcomes of the experiment. A collection $\mathcal{F}$ of subsets of $\Omega$ with the following properties (just as before) is called a *sigma-field* or *Borel field*.

*1*   1. *If a denumerable collection of sets $A_1$, $A_2$, . . . $\epsilon\mathcal{F}$ then the union*

$$\bigcup_{i=1}^{\infty} A_i \epsilon \mathcal{F}.$$

2. *The set $\Omega$ is an element of $\mathcal{F}$.*
3. *Given any set $A\epsilon\mathcal{F}$, the complementary set $\overline{A}$ is an element of $\mathcal{F}$.*

In the case of an experiment with at most a denumerable number of elementary outcomes, it was natural to take the class of all subsets of $\Omega$ as the sigma-field of events. In fact, we would have to do so if the field of events is to contain the sets consisting of single points of the space (the sample points). Since the events are to have well-defined probabilities, a probability function (such a non-negative set function is often referred to as a measure) on a sigma-field should satisfy the following conditions:

2  1. *The probability function P is well defined on all sets $A \epsilon \mathfrak{F}$ and for such A*

$$0 \leq P(A) \leq 1.$$

2. $P(\Omega) = 1.$

3. *Given any denumerable collection of disjoint sets $A_1$, $A_2$, . . . $\epsilon \mathfrak{F}$*

$$P(\bigcup_i A_i) = \sum_i P(A_i). \tag{1}$$

Property 2.3 is commonly referred to as the countable additivity or sigma-additivity of the measure $P$. If $\Omega$ is a space with a nondenumerable set of points $w$, the sigma-field of all subsets of $\Omega$ will generally be too large a field to take for $\mathfrak{F}$. For example, if $\Omega$ is the set of real numbers $w$, $0 \leq w \leq 1$, there is no probability function $P$ defined on all subsets of $\Omega$ that agrees with the ordinary notion of length on the subintervals of $\Omega$ (see [29]). It is then natural to take $\mathfrak{F}$ as a smaller subfield that is sufficiently rich to include all sets consisting of single points and on which a probability function can be defined.

*A probability space*, as before, *is a triple* $(\Omega, \mathfrak{F}, P)$ *with $\Omega$ the space of points, $\mathfrak{F}$ a sigma-field of subsets of $\Omega$ and $P$ a probability function defined on the sets of $\mathfrak{F}$.* The sets of the sigma-field $\mathfrak{F}$ are called measurable sets or events. A class of real-valued functions $X(w)$ on $\Omega$ consistent with the sigma-field is of especial interest. These are the functions $X(w)$ such that every set of the form $\{w | X(w) \leq y\}$ with $y$ real is an element of the sigma-field $\mathfrak{F}$. Such functions $X(w)$ are called *measurable functions* or *random variables*. Notice that for real precision one ought to use the phrase "measurable functions with respect to the sigma-field $\mathfrak{F}$."

Often it is natural to consider the sigma-field generated by a specific collection $\mathcal{C}$ of subsets of $\Omega$. *This sigma-field $\mathfrak{F} = \mathfrak{F}(\mathcal{C})$ is the smallest sigma-field containing the collection of sets $\mathcal{C}$.* Generally there will be several sigma-fields containing $\mathcal{C}$. As an interesting example consider $\Omega$ as the set of points in $k$-dimensional Euclidean space, that is, $k$-tuples of real numbers $w = (w_1, . . . ,w_k)$. Let $\mathcal{C}$ be the collections of subsets of $\Omega$ of the form

$$\{w | w_1 \leq x_1, . . . , w_k \leq x_k\} \tag{2}$$

with $x_1$, . . . , $x_k$ any $k$ real numbers. The sigma-field $\mathcal{B} = \mathfrak{F}(\mathcal{C})$ generated by $\mathcal{C}$ is called the *sigma-field of $k$-dimensional Borel sets.* A

*function $X(w)$ measurable with respect to the sigma-field $\mathcal{B}$ of Borel sets is called
a Borel function.* Notice that a Borel field (sigma-field) need not be the
Borel field of Borel sets. The discussion given here is also valid for a
countably infinite dimensional space.

The family of functions $X(w)$, measurable with respect to $\mathcal{F}$, has
some very interesting properties. We enumerate these properties as
follows:

3   1.  *Given any two measurable functions $X(w)$, $Y(w)$ and any two real
numbers $\alpha$ and $\beta$, the linear combination $\alpha X(w) + \beta Y(w)$ is a measurable function.*

    2.  *The limit (if it exists) of a sequence of measurable functions $X_n(w)$,
$n = 1, 2, \ldots$,*

$$\lim_{n \to \infty} X_n(w) \tag{3}$$

*is a measurable function.*

    3.  *Given $k$ measurable functions $X_1(w), \ldots, X_k(w)$ and a Borel func-
tion $Y(x_1, \ldots, x_k)$ of $k$ (real) variables, the composite function
$Y(X_1(w), \ldots, X_k(w))$ is then measurable.*

These properties of the family of measurable functions indicate that
it is closed under certain basic and natural operations. Consider the
function

$$I_A(w) = \begin{cases} 1 & \text{if } w\epsilon A \\ 0 & \text{otherwise.} \end{cases} \tag{4}$$

This is obviously a measurable function if $A\epsilon\mathcal{F}$ This function $I_A(w)$ is
called the indicator function of the set $A$. Let $A_1, \ldots, A_k$ be sets of
$\mathcal{F}$ and $\alpha_1, \ldots, \alpha_k$ any real numbers. By repeated use of property 1
of measurable functions, it is clear that

$$\sum_j \alpha_j I_{A_j}(w) \tag{5}$$

is a measurable function.

Let us now sketch the definition of the integral (if it exists) of a
measurable function or correspondingly in probabilistic language, the
expectation of a random variable. We do this first for a non-negative
measurable function $X(w)$. The sets

$$A_{k,j} = \{w | j/2^k \leq X(w) < (j + 1)/2^k\}, \tag{6}$$

with $i = 0, 1, \ldots, k(2^k - 1)$ and $k = 1, 2, \ldots$, are measurable sets. Consider the simple approximating function

$$\sum_j \frac{j}{2^k} I_{A_{k,j}}(w) = X_k(w). \tag{7}$$

Notice that $\lim\limits_{k \to \infty} X_k(w) = X(w)$. It is natural to take the integral

$$\int X_k(w) \, dP \tag{8}$$

of $X_k(w)$ as

$$\sum_j \frac{j}{2^k} P(A_{k,j}). \tag{9}$$

Clearly the sequence of values (9) is nondecreasing. *If the limit of the values (9) is finite, we say that $X(w)$ is integrable and take*

$$\int X(w) dP = \lim_{k \to \infty} \int X_k(w) \, dP. \tag{10}$$

If $X(w)$ takes both non-negative and negative values, introduce the non-negative and negative parts of $X(w)$

$$\begin{aligned} X_+(w) &= \max\,(X(w),0) \\ X_-(w) &= X_+(w) - X(w). \end{aligned} \tag{11}$$

We say that $X(w)$ *is integrable if its non-negative and negative* parts are both integrable and in that case set

$$\int X(w) \, dP = \int X_+(w) \, dP - \int X_-(w) \, dP. \tag{12}$$

The probabilistic notation for expectation of a random variable $X(w)$ is

$$EX(w) = \int X(w) \, dP. \tag{13}$$

The integral so defined has the following natural properties:

4  1. *Given any two integrable functions $X(w)$, $Y(w)$ and any two real numbers $\alpha, \beta$ the linear combination $\alpha X(w) + \beta Y(w)$ is integrable and its integral is given by*

$$\int [\alpha X(w) + \beta Y(w)] \, dP = \alpha \int X(w) \, dP + \beta \int Y(w) \, dP. \tag{14}$$

2. *Given a sequence of functions $X_n(w)$, $n = 1, 2, \ldots$, bounded in absolute value by the integrable function $Y(w)$, with limit $\lim\limits_{n \to \infty} X_n(w) = X(w)$, then*

$$\int X(w) \, dP = \lim_{n \to \infty} \int X_n(w) \, dP. \tag{15}$$

The second property of integrals referred to above can be strengthened in the following way. *Let $X_n(w)$, $n = 1, 2, \ldots$, be a sequence of integrable functions with $X(w) = \lim\limits_{n \to \infty} X_n(w)$. Let the set*

$$B_{n,a} = \{w \mid |X_n(w)| \geq a\}. \tag{16}$$

*The functions $X_n(w)$ are said to be uniformly integrable if*

$$\int_{B_{n,a}} |X_n(w)| \, dP \to 0 \tag{17}$$

*uniformly in n as $a \to \infty$. One can show that if $X_n(w)$, $n = 1, 2, \ldots$, are uniformly integrable,* then

$$\lim_{n \to \infty} \int |X(w) - X_n(w)| \, dP = 0. \tag{18}$$

Given a sigma-field $\mathfrak{F}$ and a probability measure defined on it, one often wishes to consider any subset of an $\mathfrak{F}$ set of probability zero as measurable and assign to it probability zero. This implies that one ought to consider the completed sigma-field $\mathfrak{F}_c$ generated by $\mathfrak{F}$ and the subsets of $\mathfrak{F}$ sets of measure zero. Every set of the completed sigma-field $\mathfrak{F}_c$ differs from an appropriately chosen set of $\mathfrak{F}$ by at most a subset of an $\mathfrak{F}$ set of measure zero. Naturally one takes the measure of the $\mathfrak{F}_c$ set (considering now an extension of the measure on $\mathfrak{F}$ to $\mathfrak{F}_c$) as equal to that of the $\mathfrak{F}$ set which approximates it to within a subset of an $\mathfrak{F}$ set of measure zero.

At times it will be natural to deal with measures somewhat more general than probability measures. Such a measure, $m(A)$, will satisfy all the conditions of *2* except perhaps for $0 \leq P(A) \leq 1$, $P(\Omega) = 1$. *If these are replaced by $0 \leq m(A)$, $m(\Omega) < \infty$ we call the measure a finite measure.* Here the measure of the whole space may be some non-negative number other than one. This may be further generalized by assuming $\Omega$ *can be decomposed into a countable number of disjoint sets on each of which m acts as a finite measure, where the measure of the whole space need not be finite.* Such a measure is called a *sigma-finite measure.* The measures we deal with in this book will be at most sigma-finite. The measure of the whole space $\Omega$ for such a *sigma-finite measure m* may be infinite.

*A collection $\mathcal{C}$ of subsets of $\Omega$ satisfying 1 with condition 1 replaced by the weaker requirement*

    *1′ If a finite collection of sets $A_1$, $A_2$, $\ldots$, $A_k \epsilon \mathcal{C}$*

$$\text{then the union } \bigcup_{i=1}^{k} A_i \epsilon \mathcal{C},$$

*is called a field.*

Often one may be given a non-negative set function $m$ defined on a field $\mathcal{C}$. It is of great interest to find out when $m$ can be extended to a measure defined on the sigma-field $\mathcal{F}(\mathcal{C})$ generated by the field $\mathcal{C}$. A theorem due to Carathéodory ([53] p. 87) indicates that such an extension can be carried out if $m$ already acts like a measure on the field $\mathcal{C}$, namely $m(A) \geq 0$ for $A\epsilon\mathcal{C}$, and for any countable collection of disjoint sets $A_1, A_2, \ldots, \epsilon\mathcal{C}$ with $\cup A_i\epsilon\mathcal{C}$ (notice that this assumption would not have to be made if $\mathcal{C}$ were a sigma-field since it would automatically be true) $m(\cup A_i) = \Sigma m(A_i)$. This theorem further assures us that the extension is unique on $\mathcal{F}(\mathcal{C})$ if $m$ is sigma-finite on $\mathcal{C}$. Further, any set of the sigma-field $\mathcal{F}(\mathcal{C})$ can be approximated arbitrarily well in measure by sets of the field $\mathcal{C}$ generating $\mathcal{F}(\mathcal{C})$ in the following sense. Given a set $A\epsilon\mathcal{F}(\mathcal{C})$ and any fixed $\epsilon > 0$ there is a set $B\epsilon\mathcal{C}$ such that the symmetric difference of the sets $A$ and $B$

$$A \ominus B = (A - B) \cup (B - A) \tag{19}$$

(the set of points in either $A$ or $B$ but not both) has measure

$$m(A \ominus B) < \epsilon \tag{20}$$

if $m(A)$ is finite.

Let us examine some of the remarks and results spoken of in the preceding paragraphs. First, consider taking $\Omega$ as the set of real numbers $w$, $0 < w \leq 1$. Let $\mathcal{C}$ consist of intervals of the form

$$\{w | \alpha < w \leq \beta\}, \alpha < \beta, \tag{21}$$

and sets formed by taking unions of a finite number of such intervals. It is clear that the collection of sets $\mathcal{C}$ is a field. As already remarked above the sigma-field $\mathcal{F}(\mathcal{C})$ generated by $\mathcal{C}$ is the sigma-field of one-dimensional Borel sets. Our object is to define a measure on the Borel sets that agrees with the ordinary notion of length for intervals. Every set of $\mathcal{C}$ can be represented as a union of disjoint intervals

$$\bigcup_i \{w | \alpha_i < w \leq \beta_i\}, \alpha_i < \beta_i. \tag{22}$$

Simply set

$$m(\bigcup_i \{w | \alpha_i < w \leq \beta_i\}) = \sum_i (\beta_i - \alpha_i). \tag{23}$$

One can readily see that $m$ does act like a measure on $\mathcal{C}$. It therefore can be extended uniquely to a measure on the one-dimensional Borel sets. Notice that this measure is a probability measure since $m(\Omega) = m((0,1]) = 1$. The completion of the sigma-field of Borel sets $\mathcal{F}(\mathcal{C})$, $\mathcal{F}_c(\mathcal{C})$, is commonly referred to as the sigma-field of Lebesgue sets.

The extension of $m$ from $\mathfrak{F}(\mathcal{C})$ to $\mathfrak{F}_c(\mathcal{C})$ is called Lebesgue measure. We could equally well have taken $\Omega$ as the set of all real numbers and carried out the construction given above. $\mathfrak{F}(\mathcal{C})$ would then consist of the Borel subsets of the whole real line rather than the interval $(0,1]$ and a corresponding statement would hold for $\mathfrak{F}_c(\mathcal{C})$. Notice that Lebesgue measure on the real line is a sigma-finite measure, not a finite measure.

We now make a few remarks on product spaces. Suppose $\Omega_1, \Omega_2$ are spaces of points $w_1, w_2$ respectively. Let $\mathfrak{F}_1$ and $\mathfrak{F}_2$ be sigma-fields of sets of $\Omega_1$ and $\Omega_2$ respectively with $m_1, m_2$ measures on the sigma-fields $\mathfrak{F}_1, \mathfrak{F}_2$. The space of points $w = (w_1, w_2)$ with $w_1 \epsilon \Omega_1, w_2 \epsilon \Omega_2$ is called the product space $\Omega_1 \times \Omega_2$. Consider the collection of sets of the form $\{w = (w_1, w_2) | w_1 \epsilon A_1, w_2 \epsilon A_2\} = A_1 \times A_2$ with $A_1, A_2$ any two sets of $\mathfrak{F}_1, \mathfrak{F}_2$ respectively. The sigma-field generated by sets of this collection is the product sigma-field $\mathfrak{F}_1 \times \mathfrak{F}_2$. Generate the measure $m = m_1 \times m_2$ (called the product measure generated by $m_1, m_2$) on $\mathfrak{F}_1 \times \mathfrak{F}_2$ in the following manner. Given any set $A_1 \times A_2$ with $A_1 \epsilon \mathfrak{F}_1, A_2 \epsilon \mathfrak{F}_2$, let

$$m(A_1 \times A_2) = m_1 \times m_2(A_1 \times A_2) = m_1(A_1)m_2(A_2). \qquad (24)$$

Given any finite union of disjoint sets $A_1^{(j)} \times A_2^{(j)}$, $A_1^{(j)} \epsilon \mathfrak{F}_1$, $A_2^{(j)} \epsilon \mathfrak{F}_2$, $j = 1, \ldots, n$, set

$$m( \bigcup_{j=1}^{n} (A_1^{(j)} \times A_2^{(j)})) = \sum_{j=1}^{n} m_1(A_1^{(j)})m_2(A_2^{(j)}). \qquad (25)$$

The collection of finite unions of sets of the form $A_1 \times A_2$, $A_1 \epsilon \mathfrak{F}_1$, $A_2 \epsilon \mathfrak{F}_2$, is a field and it can be shown that $m$ is countably additive on this field. One can therefore extend $m$ to a measure on the sigma-field $\mathfrak{F}_1 \times \mathfrak{F}_2$ generated by this field using Carathéodory's theorem. The same procedure can be used to construct product spaces, product fields, and product measures generated by any finite number of spaces, fields, and measures. Notice that $\mathfrak{F}_1 \times \mathfrak{F}_2$ is not a completed sigma-field with respect to $m_1 \times m_2$.

Let $f(w) = f(w_1, w_2)$ be a function measurable with respect to $\mathfrak{F}_1 \times \mathfrak{F}_2$. One can show that for each fixed $w_2$, $f(w_1, w_2)$ is measurable with respect to $\mathfrak{F}_1$ in $w_1$ and for each fixed $w_1$ measurable with respect to $\mathfrak{F}_2$ in $w_2$. Assume that $f(w)$ is integrable with respect to $m = m_1 \times m_2$ on $\Omega_1 \times \Omega_2$. A result commonly referred to as Fubini's theorem states that the integral of $f$ with respect to $m_1 \times m_2$ can be written as an iterated integral

$$\int_{\Omega_1 \times \Omega_2} f(w)m_1 \times m_2(dw) = \int_{\Omega_1} \left\{ \int_{\Omega_2} f(w_1, w_2)m_2(dw_2) \right\} m_1(dw_1)$$
$$= \int_{\Omega_2} \left\{ \int_{\Omega_1} f(w_1, w_2)m_1(dw_1) \right\} m_2(dw_2). \qquad (26)$$

The equation (26) implies that $f(w_1,w_2)$ is integrable with respect to $m_1$ for almost every point $w_2$ and is integrable with respect to $m_2$ for almost every point $w_1$. If $f(w)$ is nonnegative, equation (26) holds without the assumption of integrability. If one of the integrals is infinite, all of them are.

There are several notions of convergence for a sequence of random variables $X_n(w)$, $n = 1, 2, \ldots$, that are of interest. We say that *the sequence of random variables $X_n(w)$ converge almost everywhere or converge with probability one to the random variable $X(w)$ if*

$$X(w) = \lim_{n \to \infty} X_n(w) \qquad (27)$$

*except possibly on a set of $w$ points of probability measure zero. The random variables $X_n(w)$ are said to converge in probability to $X(w)$ as $n \to \infty$ if for every $\varepsilon > 0$*

$$\lim_{n \to \infty} P(|X_n(w) - X(w)| > \varepsilon) = 0. \qquad (28)$$

Whenever a condition is satisfied at all points outside of a set of probability or measure zero, it is referred to as a condition holding almost everywhere or almost surely.

Convergence almost everywhere implies convergence in probability. However, the converse does not hold as is indicated by the following example. Let $\Omega$ be the interval $[0,1]$ with $\mathfrak{F}$ the Lebesgue subsets of the interval and $P(\cdot)$ Lebesgue measure on these sets. Set $s_n = \sum_1^n 1/j$.

Let

$$f_n(u) = \begin{cases} 1 & \text{when } s_n \leq u < s_{n+1} \\ 0 & \text{otherwise} \end{cases} \qquad (29)$$

$0 \leq u < \infty$, and

$$X_n(w) = \sum_{j=0}^{\infty} f_n(j + w), \; 0 \leq w \leq 1. \qquad (30)$$

See the accompanying figure 4.1 for a graph of $f_n(u)$. The sequence of random variables $X_n(w)$ certainly converge in probability to $X(w) = 0$, but they do not converge almost everywhere. A third mode of convergence is that of convergence in mean square. *Let $X_1, X_2, \ldots, X$ be random variables with finite second moment. The random variables $X_n$ converge to $X$ in mean square as $n \to \infty$ if*

$$\lim_{n \to \infty} E|X_n - X|^2 \to 0. \qquad (31)$$

A simple application of the Chebyshev inequality indicates that convergence in mean square implies convergence in probability. A Cauchy convergence criterion holds for each of these modes of convergence. *If $X_n - X_m \to 0$ as $n,m \to \infty$ almost everywhere, in probability or in mean square, there is a random variable $X$ such that $X_n \to X$ in the corresponding sense as $n \to \infty$.* This result is referred to as the Riesz-Fisher theorem when one is dealing with mean square convergence. We have noted in section e of Chapter II that continuous functions on a finite closed interval can be uniformly approximated by polynomials. Actually one can show that a square integrable function on a finite interval can be approximated arbitrarily well by polynomials in mean square. Problem 10 of Chapter II indicates that a continuous function $f$ on $[-\pi,\pi]$ with $f(\pi) = f(-\pi)$ can be uniformly approximated by finite trigonometric series in $\cos kx$, $\sin kx$. Just as in the case of polynomial approximation any square integrable function on $[-\pi,\pi]$ can be approximated

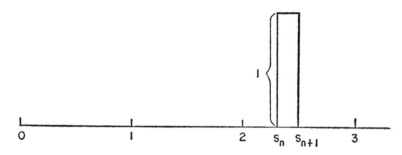

FIG. 4.1. Graph of function $f_n(u)$.

arbitrarily well in mean square by finite trigonometric series. An analogous result holds for a square integrable function of several variables on a finite interval. To be specific, suppose $f(x,y)$ is a square integrable function on $-\pi \leq x$, $y \leq \pi$. Then $f(x,y)$ can be approximated arbitrarily well in mean square by polynomials in $x$ and $y$ or by finite multiple trigonometric series in $x$ and $y$. We shall make use of this remark in deriving a representation for certain random processes in section c of Chapter VIII.

Let $X_1(w), \ldots, X_k(w)$ be $k$ random variables on the probability space $(\Omega,\mathfrak{F},P)$. Any set of the form

$$\{w|X_1(w) \leq x_1, \ldots, X_k(w) \leq x_k\} \tag{32}$$

with $x_1, \ldots, x_k$ real is an element of $\mathfrak{F}$. This implies that the function

$$F(x_1, \ldots, x_k) = P[\{w|X_1(w) \leq x_1, \ldots, X_k(w) \leq x_k\}] \tag{33}$$

is well defined for all $x_1, \ldots, x_k$. This function is called the *distribution function* of the random variables $X_1, \ldots, X_k$. Let us note the properties of such a distribution function. First of all

(i) $$0 \leq F(x_1, \ldots, x_k) \leq 1$$

(ii) $$\lim_{x_i \to -\infty} F(x_1, \ldots, x_k) = 0 \quad i = 1, \ldots, k \qquad (34)$$

(iii) $$\lim_{x_1, \ldots, x_k \to \infty} F(x_1, \ldots, x_k) = 1.$$

Furthermore $F$ is nondecreasing in each coordinate $x_i$ since the difference in the $i$-th variable

$$\Delta_{h_i} F(x_1, \ldots, x_k) = F(x_1, \ldots, x_{i-1}, x_i + h_i, x_{i+1}, \ldots, x_k) \qquad (35)$$
$$- F(x_1, \ldots, x_{i-1}, x_i, x_{i+1}, \ldots, x_k)$$
$$= P[\{w | X_j(w) \leq x_j, j = 1, \ldots, k, j \neq i, x_i < X_i(w) \leq x_i + h_i\}] \geq 0$$

for $h_i > 0$. In fact the $k$-th order difference

$$\Delta_{h_1} \Delta_{h_2} \cdots \Delta_{h_k} F(x_1, \ldots, x_k)$$
$$= P[\{w | x_i < X_j(w) \leq x_j + h_j, j = 1, \ldots, k\}] \geq 0 \qquad (36)$$

with $h_1, \ldots, h_k \geq 0$. Further $F(x_1, \ldots, x_k)$ is continuous to the right in each variable separately, that is,

$$\lim_{y_i \to x_i+} F(x_1, \ldots, x_{i-1}, y_i, x_{i+1}, \ldots, x_k)$$
$$= F(x_1, \ldots, x_{i-1}, x_i + 0, x_{i+1}, \ldots, x_k)$$
$$= F(x_1, \ldots, x_{i-1}, x_i, x_{i+1}, \ldots, x_k). \qquad (37)$$

As we have seen, a distribution function satisfies conditions (34) to (37) (notice that (36) is obviously redundant when $k = 1$). One might ask whether the converse is also true. It will be shown that any function $F(x_1, \ldots, x_k)$ satisfying conditions (34) to (37) can be considered the distribution function of an appropriately constructed set of random variables. Let $\Omega$ be the space of $k$-dimensional points $w = (w_1, \ldots, w_k)$. Consider the field $\mathcal{C}$ consisting of intervals of the form

$$I = \{w | x_i < w_i \leq x_i + h_i, i = 1, \ldots, k\}, \, h_i \geq 0, \qquad (38)$$

and all unions of a finite number of such intervals. Every set of $\mathcal{C}$ can be expressed as a union of disjoint intervals

$$\cup I_j = \cup \{w | x_i^{(j)} < w_i \leq x_i^{(j)} + h_i^{(j)}, i = 1, \ldots, k\}. \qquad (39)$$

Set

$$m(\cup I_j) = \sum_j \Delta_{h_1^{(j)}} \cdots \Delta_{h_k^{(j)}} F(x_1^{(j)}, \ldots, x_k^{(j)}).$$

The non-negative set function $m$ acts like a measure on $\mathcal{C}$ and hence can be extended to the sigma-field (the $k$-dimensional Borel sets) generated by $\mathcal{C}$ using Caratheodory's theorem. The extended measure $m$ is clearly a probability measure and the random variables $X_i(w) = w_i$, $i = 1, \ldots, k$, have $F(x_1, \ldots, x_k)$ as their joint distribution function.

Take $X_1(w), \ldots, X_k(w)$ as random variables on a probability space. If $g(x_1, \ldots, x_k)$ is a Borel measurable function of the real arguments $x_1, \ldots, x_k$ we have already remarked that $Y(w) = g(X_1(w), \ldots, X_k(w))$ is a random variable on that probability space. The expectation of $Y$, if it exists, is given by

$$EY(w) = Eg(X_1(w), \ldots, X_k(w))$$
$$= \int g(X_1(w), \ldots, X_k(w)) \, dP. \tag{40}$$

If $g$ is a continuous function of $x_1, \ldots, x_k$, the expectation can be written as a Riemann-Stieltjes integral with respect to the joint distribution function $F(x_1, \ldots, x_k)$ of the random variables $X_1(w), \ldots, X_k(w)$, just as in the discrete case,

$$EY(w) = Eg(X_1(w), \ldots, X_k(w))$$
$$= \underbrace{\int_{-\infty}^{\infty} \cdots \int g(x_1, \ldots, x_k) \, dF(x_1, \ldots, x_k).}_{k} \tag{41}$$

Often, it is more convenient, for ease in computation, to write expectations in this form.

Here are a few simple distribution functions of random variables with a continuous range of values that are of considerable interest. The first of these is the distribution function of a uniformly distributed random variable. A random variable is said to be uniformly distributed on $[a,b]$, $a < b$, if its distribution function has the form

$$F(x) = \begin{cases} 0 & \text{if } x < a \\ \dfrac{x-a}{b-a} & \text{if } a \leq x \leq b \\ 1 & \text{otherwise.} \end{cases} \tag{42}$$

Notice that the derivative $f(x)$ of $F(x)$ (called the probability density if it exists) is given by

$$f(x) = \begin{cases} \dfrac{1}{b-a} & \text{if } a < x < b \\ 0 & \text{otherwise.} \end{cases} \tag{43}$$

We encountered a more interesting class of distribution functions in our discussion of the central limit theorem (see section II.e). These are the normal distribution functions. A random variable is said to be normally distributed with mean $m$ and variance $\sigma^2 > 0$ if its distribution function is

$$F(x) = \frac{1}{\sqrt{2\pi}\,\sigma} \int_{-\infty}^{x} e^{-\frac{(u-m)^2}{2\sigma^2}}\, du. \tag{44}$$

Notice that the mean of such a random variable is

$$EX(w) = \frac{1}{\sqrt{2\pi}\,\sigma} \int_{-\infty}^{\infty} x e^{-\frac{(x-m)^2}{2\sigma^2}}\, dx = m \tag{45}$$

and its variance is given by

$$E(X(w) - m)^2 = \frac{1}{\sqrt{2\pi}\,\sigma} \int_{-\infty}^{\infty} (x-m)^2 e^{-\frac{(x-m)^2}{2\sigma^2}} = \sigma^2. \tag{46}$$

If $X(w)$ is a normally distributed random variable with mean zero and variance one, the derived random variable $Y(w) = m + \sigma X(w)$ is normally distributed with mean $m$ and variance $\sigma^2$. Of course, a random variable with mean $m$ and variance $\sigma^2 = 0$ is equal to $m$ with probability one, that is, the total probability mass is concentrated at the point $m$.

*The random variables $X_1(w), \ldots, X_k(w)$ are* called jointly *independent if*

$$P(X_1(w) \leq x_1, X_2(w) \leq x_2, \ldots, X_k(w) \leq x_k)$$
$$= P(X_1(w) \leq x_1)P(X_2(w) \leq x_2) \cdots P(X_k(w) \leq x_k) \tag{47}$$

for any set of $k$ real numbers $x_1, \ldots, x_k$. This implies the following much stronger property. Given $k$ Borel measurable functions $g_1(x_1), \ldots, g_k(x_k)$, the random variables

$$g_1(X_1(w)), \ldots, g_k(X_k(w)) \tag{48}$$

are independent if $X_1(w), \ldots, X_k(w)$ are independent. Notice that the property of independence can be immediately rewritten in terms of distribution functions

$$F(x_1, \ldots, x_k) = F_1(x_1) \cdots F_k(x_k) \tag{49}$$

where $F_1, \ldots, F_k$ are the distribution functions of the random variables $X_1, \ldots, X_k$ taken one at a time (commonly referred to as the marginal distribution functions of the random variables $X_1$,

. . . , $X_k$). Thus, if the random variables are independent, the joint distribution function of the random variables is the product of the marginal distribution functions. *An analogous property holds for the expectation of the product of independent random variables: the expectation of the product is equal to the product of the expectations of the individual random variables*

$$E[X_1(w) \cdots X_k(w)] = \int \cdots \int \prod_{i=1}^{k} x_i \, dF(x_1, \ldots, x_k)$$

$$= \int \cdots \int \prod_{i=1}^{k} x_i \, dF_1(x_1) \cdots dF_k(x_k)$$

$$= \prod_{i=1}^{k} \int x_i \, dF_i(x_i) = \prod_{i=1}^{k} E[X_i(w)]. \tag{50}$$

# b. Distribution Functions and Their Transforms

Given a random variable $X(w)$, let $\varphi(t)$ be defined as the following Fourier-Stieltjes transform of its distribution function $F(x)$

$$\varphi(t) = E \exp (itX(w)) = E[\cos (tX(w)) + i \sin (tX(w))]$$
$$= \int_{-\infty}^{\infty} e^{itx} \, dF(x). \tag{1}$$

This transform is commonly called the *characteristic function* of $X(w)$. Notice that

$$\varphi(0) = \int_{-\infty}^{\infty} dF(x) = 1 \tag{2}$$

and

$$|\varphi(t)| \leq \int_{-\infty}^{\infty} |e^{itx}| \, dF(x) = \int_{-\infty}^{\infty} dF(x) = 1. \tag{3}$$

Further $\varphi(t)$ is a continuous function of $t$. Given any finite number of $t$ points $t_1, \ldots, t_k$ and complex numbers $c_1, \ldots, c_k$ consider the quadratic form

$$\sum_{i,j=1}^{k} c_i \bar{c}_j \varphi(t_i - t_j). \tag{4}$$

The quadratic form is positive definite since

$$\sum_{i,j} c_i \bar{c}_j \varphi(t_i - t_j) = \sum_{u,v} c_u \bar{c}_v \int e^{i(t_u - t_v)x} \, dF(x) = \int \left| \sum_{u} c_u e^{it_u x} \right|^2 dF(x) \geq 0. \tag{5}$$

It is natural to ask for conditions on a function $\varphi$ ensuring that it be a characteristic function, that is, the Fourier-Stieltjes transform of a distribution function. A theorem of Bochner [53] states that the necessary conditions that $\varphi$ be continuous and (2), (5) cited above hold are also sufficient for $\varphi(t)$ to be a characteristic function.

The characteristic function of a random variable uniformly distributed on $[a,b]$ is

$$\varphi(t) = \frac{1}{b-a} \int_a^b e^{itx}\, dx = (e^{itb} - e^{ita})/(b-a). \tag{6}$$

The characteristic function of a normally distributed random variable with mean $m$ and variance $\sigma^2$ is also given readily by

$$\varphi(t) = \frac{1}{\sqrt{2\pi}\,\sigma} \int_{-\infty}^{\infty} e^{itx} e^{-\frac{(x-m)^2}{2\sigma^2}}\, dx \tag{7}$$
$$= e^{imt - \sigma^2 t^2/2}.$$

If $\varphi(t)$ is the characteristic function of the random variable $X$, the characteristic function of $aX + b$ ($a,b$ constants) is

$$E(e^{it(aX+b)}) = E(e^{itb} e^{iatX}) = e^{itb} \varphi(at). \tag{8}$$

Given random variables $X_1(w), \ldots, X_k(w)$, one can introduce a joint characteristic function $\varphi(t_1, \ldots, t_k)$ of the random variables in an analogous manner as follows

$$\varphi(t_1, \ldots, t_k) = E \exp\left( i \sum_{j=1}^{k} t_j X_j(w) \right)$$
$$= \underbrace{\int_{-\infty}^{\infty} \cdots \int}_{k} \exp\left( i \sum_{1}^{k} t_j x_j \right) dF(x_1, \ldots, x_k). \tag{9}$$

The multidimensional characteristic function satisfies conditions that parallel those mentioned in the one-dimensional case above. If $X_1(w), \ldots, X_k(w)$ are independent, the functions $\exp(it_1 X_1)$, $\ldots, \exp(it_k X_k)$ are independent. But this implies that the joint characteristic function of independent random variables $X_1, \ldots, X_k$

$$\varphi(t_1, \ldots, t_k) = E \left[ \prod_{1}^{k} \exp\left( it_j X_j \right) \right]$$
$$= \prod_{1}^{k} E \exp\left( it_j X_j \right) = \varphi_1(t_1) \cdots \varphi_k(t_k) \tag{10}$$

is equal to the product of the characteristic functions of the individual random variables $\varphi_1(t_1), \ldots, \varphi_k(t_k)$. Thus, the characteristic function of the sum $\sum_1^k X_i$ of independent random variables is the product

$$\prod_1^k \varphi_i(t). \tag{11}$$

There is a one-one correspondence between characteristic functions and corresponding distribution functions. The definition of $\varphi(t)$ indicates how a characteristic function is given in terms of the corresponding distribution function. There are inversion formulas indicating how the distribution function can be recovered from knowledge of the characteristic function. We shall encounter a few of these inversion formulas later on.

Often, it is more convenient to carry out a discussion in terms of characteristic functions rather than distribution functions. This is so in part when considering jointly normally distributed random variables. We say that *the random variables $X_1(w), \ldots, X_k(w)$ are jointly normally distributed if every linear combination of the random variables*

$$t_1 X_1(w) + \cdots + t_k X_k(w) \tag{12}$$

*is a normally distributed random variable.* Thus, a definition of $k$-dimensional normality is introduced in terms of one-dimensional normality. The means and variances of $X_1, \ldots, X_k$ clearly exist. Let them be denoted by $m_1, \ldots, m_k$ and $\sigma_1^2, \ldots, \sigma_k^2$ respectively. The mixed central moments

$$r_{u,v} = E(X_u - m_u)(X_v - m_v) \qquad u, v = 1, \ldots, k \tag{13}$$

are called the covariances of the random variables and they exist by the Schwarz inequality since

$$|r_{u,v}| \leq E|(X_u - m_u)(X_v - m_v)| = \int |(X_u(w) - m_u)(X_v(w) - m_v)| \, dP$$
$$\leq [\int |X_u(w) - m_u|^2 \, dP \int |X_v(w) - m_v|^2 \, dP]^{1/2} = \sigma_u \sigma_v. \tag{14}$$

The covariance matrix of the random variables

$$\mathbf{R} = (r_{u,v}; u, v = 1, \ldots, k) \tag{15}$$

is positive definite since

$$\sum_{u,v} c_u r_{u,v} \bar{c}_v = E[\sum_{u,v} c_u \bar{c}_v (X_u - m_u)(X_v - m_v)]$$
$$= E|\sum_u c_u (X_u - m_u)|^2 \geq 0. \tag{16}$$

Notice that the positive definiteness of the covariance matrix is generally true. It does not stem from normality. We shall now obtain the joint characteristic function of the jointly normal random variables $X_1, \ldots, X_k$. Consider the random variable

$$Y = t_1 X_1 + \cdots + t_k X_k. \tag{17}$$

By the very definition of joint normality, the random variable $Y$ is normal with mean

$$EY = t_1 m_1 + \cdots + t_k m_k = \mathbf{tm}' \tag{18}$$

and variance

$$
\begin{aligned}
E(Y - \mathbf{tm}')^2 &= E[\mathbf{t}(\mathbf{X} - \mathbf{m})']^2 \\
&= \mathbf{t}E(\mathbf{X} - \mathbf{m})'(\mathbf{X} - \mathbf{m})\mathbf{t}' \\
&= \mathbf{tRt}',
\end{aligned}
\tag{19}
$$

where $\mathbf{t}, \mathbf{m}, \mathbf{X}$ are the row vectors of $t_i$'s, $m_i$'s, and $X_i$'s respectively. The characteristic function of $Y$ is then

$$\varphi_Y(\tau) = E(e^{i\tau Y}) = \exp\{-\tfrac{1}{2}\tau^2 \mathbf{tRt}' + i\tau \mathbf{tm}'\}. \tag{20}$$

However, the joint characteristic function of $X_1, \ldots, X_k$ is given by

$$
\begin{aligned}
\varphi(t_1, \ldots, t_k) = \varphi_Y(1) = E(e^{iY}) &= E(e^{i\sum\limits_{1}^{k} t_i X_i}) \\
&= \exp\{-\tfrac{1}{2}\mathbf{tRt}' + i\mathbf{tm}'\}.
\end{aligned}
\tag{21}
$$

Of course, one would like to invert this expression if possible. Let us *assume that* $\mathbf{R}$ *is nonsingular.* This means that $\mathbf{R}$ is a positive definite matrix. The matrix $\mathbf{R}$ has a unique positive definite square root $\mathbf{R}^{1/2}$. Let $\mathbf{R}^{-1/2}$ be the inverse of $\mathbf{R}^{1/2}$. Consider the new $k$-vector $\mathbf{Z}$ of random variables $Z_i$

$$\mathbf{Z} = (\mathbf{X} - \mathbf{m})\mathbf{R}^{-1/2}. \tag{22}$$

The random variables $Z_i$ are jointly normal since they are obtained from the $X_i$'s by a linear transformation. The mean of the $\mathbf{Z}$ vector is the null vector

$$E\mathbf{Z} = E(\mathbf{X} - \mathbf{m})\mathbf{R}^{-1/2} = 0 \tag{23}$$

and the covariance matrix of the vector is

$$
\begin{aligned}
E\mathbf{Z}'\mathbf{Z} &= \mathbf{R}^{-1/2}E[(\mathbf{X} - \mathbf{m})'(\mathbf{X} - \mathbf{m})]\mathbf{R}^{-1/2} \\
&= \mathbf{R}^{-1/2}\mathbf{R}\mathbf{R}^{-1/2} = \mathbf{I}.
\end{aligned}
\tag{24}
$$

The characteristic function of the $Z_i$ random variables is thus

$$\varphi(t_1, \ldots , t_k) = \exp\{-\tfrac{1}{2}\mathbf{tt'}\} = \prod_{j=1}^{k} \exp(-\tfrac{1}{2}t_j^2). \tag{25}$$

But this is the characteristic function of $k$ independent normally distributed random variables with mean zero and variance 1. Their joint density function is therefore

$$g(z_1, \ldots , z_k) = \frac{1}{(2\pi)^{k/2}} \exp\left\{-\tfrac{1}{2}\sum_{j=1}^{k} z_j^2\right\}. \tag{26}$$

The joint density function of the $X_i$ random variables can simply be obtained by making use of the transformation

$$\mathbf{z} = (\mathbf{x} - \mathbf{m})\mathbf{R}^{-\frac{1}{2}}. \tag{27}$$

The joint density function of the $X_i$ random variables is given by

$$f(x_1, \ldots , x_k) = \frac{J}{(2\pi)^{k/2}} \exp\{-\tfrac{1}{2}\mathbf{zz'}\}$$

$$= \frac{1}{(2\pi)^{k/2}|\mathbf{R}|^{\frac{1}{2}}} \exp\{-\tfrac{1}{2}(\mathbf{x} - \mathbf{m})\mathbf{R}^{-1}(\mathbf{x} - \mathbf{m})'\} \tag{28}$$

where $J = |\mathbf{R}|^{-\frac{1}{2}}$ is the Jacobian of the transformation (27) and $|\mathbf{R}|$ denotes the determinant of $\mathbf{R}$.

    *Given a sequence of distribution functions $F_n(x)$, they are said to converge to a limiting distribution function $F(x)$ if $\lim_{n\to\infty} F_n(x) = F(x)$ for every point $x$ at which $F$ is continuous.* It should be noted that the set of points $x$ at which a distribution function $F$ is discontinuous can be at most countable since $F$ is nondecreasing (see [1] p. 85). Such limit theorems are often of considerable interest. In fact, several theorems of this type were obtained in Chapter II. If the sequence of distribution functions $F_n$ converge to a distribution function $F$, the corresponding characteristic functions

$$\varphi_n(t) = \int_{-\infty}^{\infty} e^{itx}\, dF_n(x) \tag{29}$$

converge to the limiting characteristic function

$$\varphi(t) = \int_{-\infty}^{\infty} e^{itx}\, dF(x). \tag{30}$$

A converse of this fact, commonly called the continuity theorem for characteristic function, holds and runs as follows. *If a sequence of char-*

*acteristic functions $\varphi_n(t)$ converges to a limit function $\varphi(t)$ that is continuous at zero, the limit function $\varphi(t)$ is a characteristic function and the sequence of corresponding distribution functions $F_n(x)$ converges to the distribution function $F(x)$ corresponding to the characteristic function $\varphi(t)$ at every continuity point of $F(x)$.* Thus, one can obtain limit theorems for distribution functions by proving limiting results for characteristic functions. This is exceedingly handy whenever the characteristic function domain is a more natural one to deal with. Such is the case generally when dealing with the distribution functions of sums of independent random variables. We shall show this by giving an alternative proof of the central limit theorem derived in Chapter II. The proof given here will make use of characteristic function techniques. Let $X_1, \ldots, X_n$ be $n$ independent identically distributed random variables with mean zero and variance one. Our object is to show that

$$\sum_{1}^{n} X_j / \sqrt{n} \tag{31}$$

is asymptotically normally distributed with mean zero and variance one. Let the common characteristic function of the $X_i$'s be $\varphi(t)$. The existence of the second moment implies that $\varphi$ is doubly differentiable,

$$\varphi''(t) = -\int_{-\infty}^{\infty} e^{itx} x^2 \, dF(x), \tag{32}$$

where $F(x)$ is the common distribution function of the $X_i$'s. Let us carry out a Taylor expansion of $\varphi(t)$ about $t = 0$ with error term. Then

$$\varphi(t) = 1 + \varphi'(0)t + \varphi''(0)\frac{t^2}{2} + o(t^2)$$

$$= 1 - \frac{t^2}{2} + o(t^2) \tag{33}$$

since

$$\varphi'(0) = i \int_{-\infty}^{\infty} x \, dF(x) = iEX = 0$$

$$\varphi''(0) = -\int_{-\infty}^{\infty} x^2 \, dF(x) = -EX^2 = -1. \tag{34}$$

Now the characteristic function of (31) is

$$[\varphi(t/\sqrt{n})]^n$$

by the multiplicative property (11) when dealing with independent random variables. But

$$[\varphi(t/\sqrt{n})]^n = \left(1 - \frac{t^2}{2n} + o\left(\frac{t^2}{n}\right)\right)^n \to e^{-t^2/2} \tag{35}$$

as $n \to \infty$. The limit function $\exp(-t^2/2)$ is the characteristic function of the normal distribution with mean zero and variance one. An application of the continuity theorem for characteristic functions immediately implies the central limit theorem. Of course, this proof looks simpler and more immediate than that of Chapter II. However, this is an illusion in part. First of all we are using more sophisticated techniques. Further, rather nontrivial and unproved results like the continuity theorem for characteristic functions have been employed.

Earlier in this section, inversion formulas to recover the distribution function from the corresponding characteristic function were referred to. The most famous formula of this sort is due to P. Lévy. Let $F(x)$ be a distribution function and $\varphi(t)$ the corresponding characteristic function. The P. Lévy inversion formula states that

$$F(b) - F(a) = \lim_{T \to \infty} \frac{1}{2\pi} \int_{-T}^{T} \frac{e^{-ita} - e^{-itb}}{it} \varphi(t) \, dt \qquad (36)$$

for any two continuity points $a, b$ of $F$. Since $F$ has at most a countable number of discontinuity points and is continuous to the right, this is enough to determine $F$ everywhere. If $\varphi(t)$ is integrable, $F$ is differentiable with a density function $f$

$$F(x) = \int_{-\infty}^{x} f(u) \, du \qquad (37)$$

and the density function can be given as an ordinary Fourier transform,

$$f(x) = \frac{1}{2\pi} \int_{-\infty}^{\infty} e^{-itx} \varphi(t) \, dt. \qquad (38)$$

The reader should refer to Loève [53] for a discussion of inversion formulas for multidimensional characteristic functions.

## c. Derivatives of Measures and Conditional Probabilities

One can regard a measure $m$ as an indicator of a mass distribution over the space $\Omega$, $m(A)$ with $A \in \mathfrak{F}$ the amount of mass located in the set $A$. Consider two measures $m_1, m_2$ on the sigma-field $\mathfrak{F}$. *The measure $m_1$ is said to be absolutely continuous with respect to $m_2$ if for each $A \in \mathfrak{F}$ for which $m_2(A) = 0$, $m_1(A) = 0$.* A basic result of Radon and Nikodym states that if $m_1$ is absolutely continuous with respect to $m_2$, $m_1$ has a derivative with respect to $m_2$. Specifically, the Radon-Nikodym theorem states that *if $m_1$ is absolutely continuous with respect to $m_2$, there is a function $f(w)$ measurable with respect to $\mathfrak{F}$ and uniquely defined up to a set of $m_2$ measure zero*, such that

$$m_1(A) = \int_A f(w) \, dm_2 \tag{1}$$

*for every set $A\epsilon\mathfrak{F}$*. If $\Omega$ is $k$-dimensional space with $\mathfrak{F}$ the sigma-field of $k$-dimensional Borel or Lebesgue sets and $m_2$ $k$-dimensional Lebesgue measure, it is common to simply refer to $m_1$ as absolutely continuous if it is absolutely continuous with respect to $m_2$.

In Chapter II we introduced the notion of conditional probability in the context of a probability space with a finite number of sample points. It is not obvious how this concept ought to be introduced in a probability space with an uncountable number of sample points. For suppose $X(w)$ is a random variable with a continuous range of values on the probability space. Given a set $A\epsilon\mathfrak{F}$, one would like to make the conditional probability of $A$ given $X(w) = x, P(A|X(w) = x)$, meaningful. We call the sigma-field generated by the family of sets $\{w|X(w) \leq x\}$ where $x$ is any real number, the sigma-field $\mathfrak{F}_X$ generated by the random variable $X(w)$. The sigma-field $\mathfrak{F}_X$ generated by the random variable $X(w)$ is a subsigma-field of $\mathfrak{F}$. If the conditional probability $P(A|X(w) = x)$ is to be thought of for fixed $A$ as a function on $\Omega$, it is natural to take it as a function measurable with respect to the subsigma-field $\mathfrak{F}_X$. In any case this suggests that we rephrase the problem as one of defining the conditional probability of an event given a sigma-field rather than that of the conditional probability of an event given a value of a random variable since the second can be subsumed as a special case of the first

$$P(A|X(w) = x) = P(A|\mathfrak{F}_X)(w). \tag{2}$$

In Chapter II, it was noted that the conditional probability function satisfied relation (II.b.3). An analogue of this relation will be used as the definition of the conditional probability function in general. The Radon-Nikodym theorem will be heavily used in this definition.

Suppose we wish to define the conditional probability of $A\epsilon\mathfrak{F}$ given the subsigma-field $\mathfrak{B}$ of $\mathfrak{F}$, $P(A|\mathfrak{B})(w)$. Consider the two measures $P(AB) = P_A(B)$ and $P(B)$ on the sets $B$ of the sigma-field $\mathfrak{B}$. Since $0 \leq P(AB) \leq P(B)$, it follows that $P_A$ is absolutely continuous with respect to $P$ on $\mathfrak{B}$. There is then a derivative $f_A(w)$ of $P_A$ with respect to $P$ on $\mathfrak{B}$ that is measurable with respect to $\mathfrak{B}$ and uniquely defined up to an $w$ set of $P$ measure zero. *This derivative $f_A(w)$ is characterized by the fact that it satisfies the relation*

$$P_A(B) = P(AB) = \int_B f_A(w)\, dP \qquad (3)$$

*for every $B \epsilon \mathcal{B}$ and is measurable with respect to $\mathcal{B}$. The conditional probability of $A$ given the sigma-field $\mathcal{B}$ is defined to be $f_A(w)$,*

$$P(A|\mathcal{B})(w) = f_A(w). \qquad (4)$$

Thus the conditional probability $P(A|\mathcal{B})(w)$ is determined by the relation

$$P_A(B) = P(AB) = \int_B P(A|\mathcal{B})(w)\, dP,\; B \epsilon \mathcal{B}, \qquad (5)$$

and its $\mathcal{B}$ measurability as a function of $w$. The relation (5) implies that

(i) $\qquad\qquad\qquad 0 \leq P(A|\mathcal{B})(w) \leq 1 \qquad\qquad\qquad (6)$
(ii) $\qquad\qquad\qquad P(\Omega|\mathcal{B})(w) = 1 \qquad\qquad\qquad (7)$

(iii) for a countable collection $A_1, A_2, \ldots$ of disjoint sets of $\mathcal{F}$

$$P(\cup\, A_i|\mathcal{B})(w) = \sum_i P(A_i|\mathcal{B})(w) \qquad (8)$$

except for a $w$ set of $P$ measure zero. Notice that these relations imply that one ought to be able to treat the conditional probability function $P(A|\mathcal{B})(w)$ for fixed $w$ as a measure in the sets $A \epsilon \mathcal{F}$. There is only one difficulty. In dealing with a property like (iii) of the conditional probability function, there is an exceptional $w$ set of $P$ measure zero on which the relation may not hold. The exceptional $w$ set corresponds to the countable collection $A_1, A_2, \ldots$ . Since there may be an uncountable number of such countable collections $A_i$, we may be confronted with an uncountable number of exceptional $w$ sets, each of $P$ measure zero, that add up to a $w$ set of positive $P$ measure. This would be exceedingly unpleasant. At the worst, one would like to have an exceptional $w$ set of $P$ measure zero that is the same for all sets $A \epsilon \mathcal{F}$. Even though this is not true generally, it can be made to be the case in the contexts of greatest interest. A detailed discussion of this question can be found in Doob's book [12]. Whenever a use of the conditional probability function as a measure in $A$ is required in our discussions, we will be in a context where it can be arranged. Thus, whenever the conditional probability function is spoken of as a measure, it will satisfy the following requirements: *(a) For each $A \epsilon \mathcal{F}$, as a function of $w$ $P(A|\mathcal{B})(w)$ is* *measurable with respect to* $\mathcal{B}$. (*b*) *For each* $w \epsilon \Omega$ (notice that this is apparently stronger than "for almost every $w$" but not essentially) $P(A|\mathcal{B})(w)$ *is a measure on sets* $A \epsilon \mathcal{F}$. Of course, it must be a function satisfying these requirements which is consistent with relation (5). The conditional expectation (or mean) of one random variable $Y(w)$ given another random variable $X(w)$, or more appropriately given the sigma-field $\mathcal{F}_X$ generated by $X(w)$, is now plausibly given by

$$E[Y|X(w) = x] = E[Y|\mathcal{F}_X](w) = \int Y(w')P(dw'|\mathcal{F}_X)(w) \qquad (9)$$

the integral of $Y(w)$ with respect to the measure $P(A|\mathcal{F}_X)(w)$. Thus $E[Y|\mathcal{F}_X](w)$ is a function that is $\mathcal{F}_X$ measurable. More generally consider the conditional expectation of $Y$ with respect to a sigma-field $\mathcal{B}$ (a subsigma-field of $\mathcal{F}$) not necessarily generated by a single random variable $X$. Then $E(Y|\mathcal{B})(w)$ is the integral of $Y$ with respect to the measure $P(A|\mathcal{B})(w)$

$$E(Y|\mathcal{B})(w) = \int Y(w')P(dw'|\mathcal{B})(w). \qquad (10)$$

Notice that relations (5), (10) suggest that the conditional expectation $E(Y|\mathcal{B})(w)$ satisfies

$$\int_B Y(w)\, dP = \iint_B Y(w')P(dw'|\mathcal{B})(w)\, dP$$

$$= \int_B E(Y|\mathcal{B})(w)\, dP \qquad (11)$$

for every $B \epsilon \mathcal{B}$. This last equality is sometimes taken as the defining relation for $E(Y|\mathcal{B})(w)$ together with the condition that the conditional expectation be $\mathcal{B}$ measurable. Notice that the conditional probability $P(A|\mathcal{B})(w)$ can be considered as a special case of a conditional expectation with $Y(w)$ taken as the characteristic function $c_A(w)$ of the set $A$

$$P(A|\mathcal{B})(w) = E(c_A|\mathcal{B})(w).$$

We mention an important property of conditional expected values. Let $\mathcal{B}'$, $\mathcal{B}$ be two subsigma-fields of $\mathcal{F}$ with $\mathcal{B}' \subset \mathcal{B}$. Then if $f$ is integrable

$$E(E(f|\mathcal{B})|\mathcal{B}') = E(f|\mathcal{B}'). \qquad (12)$$

As an example let us consider the joint distribution function $F(x_1, x_2)$ of two random variables $X_1$, $X_2$ and the probability measure $P$ on the two-dimensional Borel sets generated by the distribution function $F(x_1, x_2)$. The space $\Omega$ consists of two-dimensional points $w =$

$(w_1,w_2)$ with the first component corresponding to a possible value of $X_1$ and the second component a possible value of $X_2$. The sigma-field $\mathfrak{F}$ is the sigma-field of two-dimensional Borel sets. Let $A\epsilon\mathfrak{F}$. Suppose we wish to compute the conditional probability of $A$ given $X_2$, $P(A|X_2(w) = w_2) = P(A|\mathfrak{F}_{X_2})(w) = P(A|\mathfrak{B})(w)$. Here $\mathfrak{B} = \mathfrak{F}_{X_2}$. Thus $\mathfrak{B}$ is that subsigma-field of $\mathfrak{F}$ consisting of sets with no restriction on the first component $w_1$ of $w$. Thus $P(A|\mathfrak{B})(w)$ depends on $w$ only through the second component $w_2$ as is to be expected, due to its measurability with respect to $\mathfrak{B}$. Suppose the measure $P$ is absolutely continuous with respect to two-dimensional Lebesgue measure. Then, by the Radon-Nikodym theorem, there is a density function $f(w_1,w_2) \geq 0$ such that

$$P(A) = \iint_A f(w_1,w_2)\ dw_1\ dw_2. \tag{13}$$

The probability measure on the sets $B\epsilon\mathfrak{B} = \mathfrak{F}_{X_2}$ is given by

$$P(B) = \int_B \int_{-\infty}^{\infty} f(w_1,w_2)\ dw_1\ dw_2. \tag{14}$$

Assuming $f(w_1,w_2) > 0$ everywhere, the conditional probability $P(A|\mathfrak{B})(w)$ is given by

$$P(A|\mathfrak{B})(w) = \frac{\displaystyle\int_{\{w_1|(w_1,w_2)\,\epsilon A\}} f(w_1,w_2)\ dw_1}{\displaystyle\int_{-\infty}^{\infty} f(w_1,w_2)\ dw_1}. \tag{15}$$

Notice that

$$f(w_1,w_2) \Big/ \int_{-\infty}^{\infty} f(w_1,w_2)\ dw_1 \tag{16}$$

is the conditional density function, that is, the density function of the conditional probability measure. The discussion of the diametrically opposite case in which the measure $P$ has a countable number of points $w$ on which all the probability mass is concentrated parallels that given in section b of Chapter II.

Suppose we illustrate the example of the previous paragraph by taking the two random variables $X_1,X_2$ jointly normally distributed with nonsingular covariance matrix $\mathbf{R}$ and vector of means $\mathbf{m}$. The joint density function $f(w_1,w_2)$ of $X_1,X_2$ is given by

$$f(w_1,w_2) = (2\pi)^{-1}|\mathbf{R}|^{-\frac{1}{2}} \exp\ \{-\tfrac{1}{2}(\mathbf{w} - \mathbf{m})\mathbf{R}^{-1}(\mathbf{w} - \mathbf{m})'\} \tag{17}$$

and the marginal density function of $X_2$ alone is

$$f_2(w_2) = (2\pi r_{22})^{-\frac{1}{2}} \exp\ \{-(w_2 - m_2)^2/(2r_{22})\}. \tag{18}$$

The conditional density function of $X_1$ given $X_2$ becomes

$$f(w_1|w_2) = f(w_1,w_2)/f_2(w_2) \tag{19}$$

$$= [2\pi(r_{11} - r_{12}^2/r_{22})]^{-\frac{1}{2}} \exp\left\{-\frac{r_{22}}{2(r_{11}r_{22} - r_{12}^2)}\left[w_1 - m_1 - \frac{r_{12}}{r_{22}}(w_2 - m_2)\right]^2\right\}.$$

It is clear that the conditional distribution of $X_1$ given $X_2$ is itself a normal distribution with mean

$$m_1 + \frac{r_{12}}{r_{22}}(w_2 - m_2) \tag{20}$$

and variance

$$r_{11} - r_{12}^2/r_{22}. \tag{21}$$

## d. Random Processes

*A random or stochastic process* $\{X_t(w), t \epsilon T\}$ *is an indexed family of random variables* $X_t(w)$ *on a probability space with the index* $t$ *ranging over some parameter set* $T$. Of course, the implicitly given probability measure determines the joint probability structure of the random variables $X_t$. If the process is to represent some aspect of a system subject to random fluctuations through time, the parameter $t$ will typically be thought of as time. For a system observed at discrete time points, the index set $T$ might be the set of lattice points $\{kh; k = 0, \pm 1, \ldots\}$, $h > 0$. If observation is continuous, $T$ is the set of real numbers. Such would be the case if we were observing random fluctuations of current through a cable or the fluctuating water level behind a dam through time. An example of a multidimensional index $t$ is given by turbulent fluid motion. $X_t(w)$ can be taken as one component of the random velocity of the fluid at location $t$. Here $t$ is three dimensional. Another simple example of a stochastic process with a multidimensional parameter $t$ is given by

$$X_t = \sum_j \cos(t\lambda_j')Z_j \tag{1}$$

where the $Z_j$'s are a finite number of independent normal variables and $t$ and the $\lambda_j$'s are $k$-vectors.

In section a of this chapter, given a distribution function

$$F(x_1, \ldots, x_k)$$

we noted how a probability measure $P$ could be set up on the space of $k$-dimensional points $w = (w_1, \ldots, w_k)$ so that the coordinates $X_i(w) = w_i$, $i = 1, \ldots, k$, are random variables with $F(x_1, \ldots, x_k)$ as their distribution function. This is a rather natural construction since the points $w$ of the probability space can be thought of as sample points. A corresponding question arises when we consider a random process, generally a collection of an infinite number (possibly uncountable) of random variables.

Given a stochastic process $\{X_t(\omega), t\epsilon T\}$, the distribution function

$$F^{(X_{t_1}, \ldots, X_{t_k})}(x_{t_1}, \ldots, x_{t_k}) \tag{2}$$

of any finite number of random variables $X_{t_1}, \ldots, X_{t_k}, t_1, \ldots, t_k \epsilon T$, is determined. This family of the joint distribution functions of any finite collection of random variables of the process satisfies certain obvious consistency conditions. Suppose $Q$ is a permutation of the integers $1, \ldots, k$ among themselves. Then clearly

$$F^{(X_{t_{Q1}}, \ldots, X_{t_{Qk}})}(x_{t_{Q1}}, \ldots, x_{t_{Qk}}) = F^{(X_{t_1}, \ldots, X_{t_k})}(x_{t_1}, \ldots, x_{t_k}) \tag{3}$$

since the order in which the random variables are listed is irrelevant. Further, if we let the variables $x_{t_{j+1}}, \ldots, x_{t_k}$, $1 \leq j < k$, approach infinity, the distribution function of the random variables $X_{t_1}, \ldots, X_{t_j}$ is obtained

$$F^{(X_{t_1}, \ldots, X_{t_k})}(x_1, \ldots, x_j, \infty, \ldots, \infty)$$
$$= F^{(X_{t_1}, \ldots, X_{t_j})}(x_1, \ldots, x_j), \quad 1 \leq j < k. \tag{4}$$

Typically in a description of a random process, the measure space and the probability measure on the space are not given. One simply describes the family of joint distribution functions of every finite collection of random variables of the process. The question that arises in the case of such a specification is as to whether there actually is a random process with such a family of joint distribution functions. A theorem of Kolmogorov [44] assures us that this is the case if the joint distribution functions satisfy the natural consistency conditions (3), (4). The probability space on which the process is constructed has functions $w = (w(t), t\epsilon T)$ as its "sample points." Thus a sample point $w = (w(t), t\epsilon T)$ corresponds to a possible realization of the whole process. Since a function $(w(t), t\epsilon T)$ can really be thought of as a vector (generally infinite-dimensional) with dimension the number of index points in $T$, this is an appropriate generalization of the finite dimensional probability space spoken of above. The random variable $X_t(w)$ is simply the value of $w = (w(t), t\epsilon T)$ at the fixed point $t$

$$X_t(w) = w(t). \tag{5}$$

The sigma-field $\mathfrak{F}$ that the probability measure $P$ is defined on, is generated by the sets of the form

$$\{w|w(t_1) \leq x_1, \ldots, w(t_k) \leq x_k; t_1, \ldots, t_k \epsilon T\}. \qquad (6)$$

In effect, a set function is defined on the field of sets generated by those of the form (6), starting from the given joint distribution functions. It is then shown that the set function actually acts like a measure on the field and hence by Caratheodory's theorem can be extended to a measure on the sigma-field $\mathfrak{F}$ generated by the field.

The sigma-field $\mathfrak{F}$ considered in the Kolmogorov theorem is rich enough to contain events of the form

$$\{w|w(t_1) \leq x_1, \ldots, w(t_k) \leq x_k, \ldots; t_1, t_2, \ldots \epsilon T\}$$
$$= \{w|X_{t_1}(w) \leq x_1, \ldots, X_{t_k}(w) \leq x_k, \ldots; t_1, t_2, \ldots \epsilon T\}, \qquad (7)$$

that is, events characterized by conditions imposed on a countable number of the random variables $X_t$. However, in many contexts, there is an interest in events of the form

$$\{w|\max_{t \epsilon U}|X_t(w)| \leq a\} \qquad (8)$$

where $U$ is a subset of $T$. If $U$ contains an uncountable number of points, an event of this type will not be in $\mathfrak{F}$. However if the parameter set $T$ has a topology on it (a neighborhood notion) and the properties of $\{X_t, t \epsilon T\}$ are decent as $t$ varies through $T$, we may hope to be able to set up the probability space on which the process is defined in such a way that the difference between (8) and

$$\{w|\max_{t \epsilon U_c}|X_t(w)| \leq a\} \qquad (9)$$

is a set of probability zero, when $U_c$ is a countable subset of $U$ with points that are dense in $U$. Then a set of type (8) specified by conditions at an uncountable number of points $t$ is in $\mathfrak{F}$. Whenever this is possible, it follows from the fact that the space of sample points $w = \{w(t), t \epsilon T\}$ need not be taken as the set of all functions on $T$; it can be taken as a set of functions with some strong regularity properties such as continuity, piece-wise continuity, or bounded variation depending on the joint probability distribution of the random variables $X_t$. Then the form of a sample function can be determined by its behavior on a dense

set of points. Whenever an event specified by conditions on an uncountable number of random variables arises in a specific context, it will be a
context in which it can be phrased in terms of a countable number of
conditions on an appropriate probability space. A discussion of conditions under which such a replacement can be effected can be found in
Doob [12] where he speaks of them as "separability" conditions.

Let us now consider a few continuous parameter random processes.
The first of these is the simplest normal or Gaussian process. *A process
is called a normal process if the joint distribution of any finite number of the
random variables is normal.* This process is usually called the "Wiener" or
"Brownian motion" process. It has been used as a crude model of a
particle in Brownian motion. The particle starts out from zero at time
zero

$$X_0(w) \equiv 0. \tag{10}$$

The displacements of the particle over nonoverlapping time intervals
are independent and normal with means zero and variances equal to the
lengths of the intervals. Thus

$$X_{t_i}(w) - X_{\tau_i}(w) \qquad i = 1, \ldots, k \tag{11}$$

$0 \leq \tau_1 \leq t_1 \leq \cdots \leq \tau_k \leq t_k$ are independent with means zero and
variances $t_i - \tau_i$ respectively. Such a process is sometimes referred to as
a process with independent increments. This implies that the joint
probability density of the random variables $X_{t_1}(w), \ldots, X_{t_k}(w)$ is
given by

$$\frac{1}{\sqrt{2\pi t_1}} \exp\left(-\frac{x_1{}^2}{2t_1}\right) \prod_{j=1}^{k-1} \frac{1}{\sqrt{2\pi(t_{j+1} - t_j)}} \exp\left(-\frac{(x_{j+1} - x_j)^2}{2(t_{j+1} - t_j)}\right). \tag{12}$$

The sample functions (points) of this process can be taken to be continuous (see [12]).

A second example of a continuous time parameter process with
independent increments is given by the Poisson process. Here, as
before,

$$X_0(w) \equiv 0. \tag{13}$$

Now the increments of the process over nonoverlapping time intervals
are independent Poisson variables with means equal to $\lambda(>0)$ multi-

plied by the lengths of the time intervals. This implies that the probability

$$P(X_{t_1}(w) = j_1, \ldots, X_{t_k}(w) = j_k; 0 < t_1 < \cdots < t_k)$$

$$= \begin{cases} \dfrac{(\lambda t_1)^{j_1} e^{-\lambda t_1}}{j_1!} \prod_{u=1}^{k-1} \dfrac{[\lambda(t_{u+1} - t_u)]^{j_{u+1}-j_u}}{(j_{u+1} - j_u)!} e^{-\lambda(t_{u+1}-t_u)} \\ \quad \text{if the integers } j_1, \ldots, j_k \text{ satisfy } 0 \le j_1 \le \cdots \le j_k \quad (14) \\ 0 \text{ otherwise.} \end{cases}$$

This process is sometimes used in a simplified model of a telephone exchange. $X_t(w)$ then represents the number of calls made from time 0 to time $t$. The sample functions of this process can be taken as jump functions (see [12]).

When dealing with a continuous time parameter random process $X_t(w)$, $0 \le t \le T$, one often assumes that $X_t(w)$ is jointly measurable in $t$ and $w$. We shall clarify what is meant by this statement. The process $X_t(w)$ is considered as a function of $(t,w)$ on the product space $[0,T] \times \Omega$ of the real line segment $[0,T]$ and the space $\Omega$ on which the random variables $X_t$ are defined. The product sigma-field generated by the sigma-field of Lebesgue measurable sets on $[0,T]$ and the sigma-field $\mathfrak{F}$ on $\Omega$ is considered. By joint measurability of $X_t(w)$ in $(t,w)$ one means measurability with respect to this product sigma-field. This assumption is a convenient one to make (if possible) in certain contexts. Let $m \times P$ be the product measure with $m$ Lebesgue measure on $[0,T]$ and $P$ the probability measure of the process on $\Omega$. Suppose that $X_t(w)$ is jointly measurable in $t$ and $w$ with a bounded second moment

$$E|X_t(w)|^2 = \int_\Omega |X_t(w)|^2 \, dP \le M < \infty, 0 \le t \le T. \qquad (15)$$

The iterated integral

$$\int_0^T E|X_t(w)|^2 \, dt \le MT \qquad (16)$$

is finite. By Fubini's theorem

$$\int_0^T E|X_t(w)|^2 \, dt = \int_{[0,T] \times \Omega} |X_t(w)|^2 m \times P(d(t,w))$$

$$= \int_\Omega \int_0^T |X_t(w)|^2 \, dt \, dP \qquad (17)$$

and hence

$$\int_0^T |X_t(w)|^2 \, dt$$

is finite for almost every $w$. The assumption of joint measurability in $t$ and $w$ is made in section c of Chapter VIII.

We have taken random variables as real-valued measurable functions with respect to the sigma-field of the probability space. In the discussion of certain types of random phenomena it is convenient to enlarge the concept of random variable. A random or stochastic process $\{X_t(w), t \epsilon T\}$ is still an indexed family of random variables $X_t(w)$ on a probability space with index $t$ ranging over some parameter set $T$ but the random variables are allowed to be complex-valued or vector-valued (possibly infinite dimensional). If the random variables are complex-valued, the real and imaginary parts of the random variables must be measurable with respect to the sigma-field of the probability space. If the random variables are vector-valued (let us say with real-valued components), all the components of the random variables must be measurable with respect to the sigma-field of the probability space. Occasionally such an enlarged notion of random variable and random process will be allowed in the chapters following.

## e. Problems

**1.** Let $X_1, \ldots, X_n$ be $n$ independent, identically distributed random variables with an exponential distribution

$$F(x) = \begin{cases} 0 & x < 0 \\ 1 - e^{-x} & x \geq 0. \end{cases}$$

Find the distribution function of the sum $X_1 + \cdots + X_n$.

**2.** Let $X_1, \ldots, X_n$ be independent, identically distributed random variables with common density function

$$f_\mu(x) = \frac{1}{\pi} \frac{1}{1 + (x - \mu)^2}.$$

Find the probability distribution of $S_n = \frac{1}{n} \sum_{i=1}^{n} X_i$.

What does this imply about $S_n$ as an estimate of $\mu$?
Can you suggest other estimates of $\mu$? The density function $f_\mu$ is sometimes called the Cauchy density function.

**3.** Let $X$ be a random variable with a continuous distribution function $F(x)$. Find the probability distribution of $F(X)$.

**4.** What can be said about the previous example if $F(x)$ has discontinuities?

**5.** Let $X_1$, $X_2$ be two random variables with joint distribution function $F(x_1,x_2)$. Suppose that $F(x_1,x_2)$ is absolutely continuous, that is,

$$F(x_1,x_2) = \int_{-\infty}^{x_1} \int_{-\infty}^{x_2} f(y_1,y_2)\, dy_2\, dy_1.$$

Let $F_1(x_1)$ be the marginal distribution function of $X_1$ and $F(x_2|x_1)$ the conditional distribution function of $X_2$ given $X_1$. Find the joint distribution of $F_1(X_1)$ and $F(X_2|X_1)$.

**6.** Let $X_1, \ldots, X_n$ be $n$ independent random variables with common density function

$$f_\theta(x) = \begin{cases} \theta^{-1} \exp(-x/\theta) & x > 0 \\ 0 & x \leq 0. \end{cases}$$

Let $Y_1, \ldots, Y_n$, $Y_1 \leq \cdots \leq Y_n$, be the magnitudes $X_1, \ldots, X_n$ relabeled in order of size. Find the joint distribution of $Y_1, \ldots, Y_r$, $1 \leq r \leq n$.

**7.** Prove Lyapunov's form of the central limit theorem (see Problem 15 of Chapter II) by using characteristic function techniques.

**8.** Show that if a random variable $X$ has a finite positive integral absolute moment $E|X|^n < \infty$, then the characteristic function of its distribution is differentiable up to $n$-th order.

**9.** Show that if the characteristic function of the distribution of $X$ is differentiable up to even order $2n$ then

$$P[|X| \geq t] = o(t^{-2n}) \text{ as } t \to \infty.$$

**10.** Let $X_1$, $X_2$, $X_3$, $X_4$ be jointly normal random variables with common mean zero and covariances $r_{i,j} = EX_iX_j$, $i, j = 1, 2, 3, 4$. Show that the fourth-order moment $EX_1X_2X_3X_4 = r_{12}r_{34} + r_{13}r_{24} + r_{14}r_{23}$.

**11.** Let $X_1$, $X_2$ be two random variables with a given joint distribution. $X_1$ is assumed to have finite second moment. We wish to predict $X_1$ by a predictor $p(X_2)$ in terms of $X_2$. Show that the predictor of $X_1$ best in terms of minimizing the mean square error of prediction is $p(X_2) = E(X_1|X_2)$. The result can be obtained by making use of the identity $E(X_1 - p(X_2))^2 = E(E((X_1 - p(X_2))^2|X_2))$.

**12.** Apply the result of the previous example when $X_1$, $X_2$ are jointly normally distributed.

**13.** Let $f(u_1, \ldots, u_k)$ be a non-negative continuous function defined on the unit hypercube $0 \leq u_1, u_2, \ldots, u_k \leq 1$ and bounded above by one. Consider $(X_1^{(i)}, \ldots, X_k^{(i)})$, $i = 1, \ldots, n$, as independent, identically distributed points, each uniformly distributed over the $k$-dimensional hypercube. Find the mean and variance of the statistic

$$\frac{1}{n} \sum_{i=1}^{n} f(X_1^{(i)}, \ldots, X_k^{(i)}).$$

**14.** Show how the statistic introduced in the previous example can be used as an estimate of the integral

$$\int_0^1 \cdots \int f(u_1, \ldots, u_k) \, du_1 \cdots du_k.$$

This is a simple example of a Monte Carlo technique. Can you show what advantage there is in using such an estimate when $k$ is not small?

**15.** By using a table of random numbers get an estimate of

$$\int_0^1 \cdots \int \prod_{i=1}^{4} (1 - u_i/2) \, du_1 \, du_2 \, du_3 \, du_4$$

on the basis of a sample of size $n = 15$. How does this compare with the actual value?

# Notes

1. There are several ways of developing the theory of the integral. The brief heuristic discussion in section a is consistent with only one of these. It is worthwhile looking at [29], [53], [63] for alternative developments.

2. Characteristic functions (Fourier-Stieltjes transforms of distribution functions) are a basic tool in the analysis of possible limiting probability distributions of normed sums of independent random variables. The derivation of the central limit theorem for sums of identically distributed random variables with finite second moment given in section b is the simplest such application. A detailed analysis of the limiting distributions of normed sums of independent random variables can be found in Gnedenko and Kolmogorov [23]. Our principal interest is in the charac-

terization of a positive definite function $\varphi(t)$, that is, the characterization of a function satisfying (b.5) as the Fourier-Stieltjes transform of a bounded nondecreasing function. This result is applied in Chapter VII on weakly stationary processes.

3. The statistical model of turbulent fluid motion casually referred to at the beginning of section d is examined in considerable detail in Batchelor's monograph on homogeneous turbulence [2]. Other examples of random processes with a multi-dimensional parameter are encountered in statistical models of storm-generated ocean waves [60].

# V

## STATIONARY PROCESSES

### a. Definition

Consider a random process $X_t(w)$ indexed by a parameter $t \epsilon T$ for which an additive group operation is defined. Typical examples are those in which $T$ is the set of real numbers or the set of lattice points $t = 0, \pm 1, \ldots$ on the real line. More generally $T$ might be all the points in $k$-dimensional Euclidian space or the lattice points in such a space. We shall say that *the process is strictly stationary if the random variables*

$$X_{t_1}, X_{t_2}, \ldots, X_{t_m} \tag{1}$$

*have the same joint probability distribution as the random variables*

$$X_{t_1+h}, \ldots, X_{t_m+h} \tag{2}$$

*for any positive integer m, any $t_1, \ldots, t_m$ and all h in T.* This can be phrased more succinctly by stating that the probability structure of the process is invariant under parameter translation. The finite dimensional joint probability distributions therefore depend on $t_1, \ldots, t_m$ only through the differences $t_2 - t_1, \ldots, t_m - t_1$. It is worthwhile giving a few examples of stationary processes as they arise as models of various types of natural phenomena. The examples we give are set in an engineering or physics context. This is natural since it is here that stationary processes have been thought of as natural models of natural phenomena most frequently.

1. The first example is that of *shot noise*. Consider the output current $X_t$ of a vacuum tube. The current observed at time $t$ is a summation of the contributions due to electrons arriving at the anode of the tube at time $t$ or earlier. The tube and circuit can be characterized by a function $g(t)$ that gives the contribution to the current observed at time $t$ due to the arrival of an electron at time 0. Assume that the electron arrivals are independent of each other. Further let the probability of an arrival in $(t, t + \Delta t)$ be $\beta \Delta t$. Then the number $n(t)$ of arrivals of electrons during the time $(0,t)$ is a Poisson process. Assume that the

effects of the electrons superimpose linearly. The current at time $t$ is then given by

$$X_t = \sum_{t_\nu \leq t} g(t - t_\nu) \tag{3}$$

where the $t_\nu$ are the time points of arrival of the electrons. Notice that the process $X_t$ can also be written

$$X_t = \int_{-\infty}^{\infty} g(t - \tau) \, dn \, (\tau) = \int_{-\infty}^{t} g(t - \tau) \, dn \, (\tau) \tag{4}$$

since $g(t)$ would be zero for negative $t$.

2. A second example arises in statistical mechanics. Consider a conservative dynamical system with $n$ degrees of freedom [22]. The system is described by $n$ generalized coordinates $q_1, q_2, \ldots, q_n$ and the corresponding generalized momenta $p_1, p_2, \ldots, p_n$. The motion of the system is prescribed by the system of differential equations

$$\left. \begin{array}{l} \dfrac{dq_i}{dt} = \dfrac{\partial H}{\partial p_i} \\[2mm] \dfrac{dp_i}{dt} = -\dfrac{\partial H}{\partial q_i} \end{array} \right\} i = 1, \ldots, n \tag{5}$$

where $t$ is time and $H = H(p_i, q_i; i = 1, \ldots, n)$ is the Hamiltonian of the system. If $H$ is a sufficiently smooth function, the system of differential equations has a unique solution for prescribed initial values of the $p_i$'s and $q_i$'s. The $2n$-space of points $A = (x_1, x_2, \ldots, x_{2n})$, with $x_1 = q_1$, $\ldots, x_n = q_n, x_{n+1} = p_1, \ldots, x_{2n} = p_n$, is called the *phase space* of the system. A system characterized by a point $A_s$ in phase space at time $s$ is characterized by the point $A_{s+t}$ after time $t$. We have a one-parameter family of transformations $T_t$ of the phase space into itself. Consider a set $S$ of finite volume in the phase space. The transformation $T_t$ transforms the set $S$ into a set $S_t = T_t S$ with volume

$$\int_{S_t} dx_1 \cdots dx_{2n} = \int_S J \, dy_1 \cdots dy_{2n} \tag{6}$$

where $T_t(y_1, \ldots, y_{2n}) = (x_1, \ldots, x_{2n})$ and $J$ is the corresponding Jacobian

$$J = \frac{\partial(x_1, \ldots, x_{2n})}{\partial(y_1, \ldots, y_{2n})}. \tag{7}$$

We shall show that volume is preserved under these transformations. Now

$$\frac{\partial J}{\partial t} = \sum_{k=1}^{2n} J_k \tag{8}$$

where

$$J_k = \frac{\partial(x_1, \ldots, x_{k-1}, x, x_{k+1}, \ldots, x_{2n})}{\partial(y_1, \ldots, y_{2n})} \tag{9}$$

and $x = dx_k/dt$. But

$$J_k = \sum_{\nu=1}^{2n} \frac{\partial x}{\partial x_\nu} \frac{\partial(x_1, \ldots, x_{k-1}, x_\nu, x_{k+1}, \ldots, x_{2n})}{\partial(y_1, \ldots, y_{2n})} = \frac{\partial x}{\partial x_k} J \tag{10}$$

and

$$\frac{\partial J}{\partial t} = J \sum_{k=1}^{n} \frac{\partial^2 H}{\partial p_k \partial q_k} - J \sum_{k=1}^{n} \frac{\partial^2 H}{\partial q_k \partial p_k} = 0. \tag{11}$$

Thus $J$ does not depend on $t$ and therefore $J \equiv 1$. The volume is left unchanged by $T_t$. This result is known as *Liouville's theorem*. Consider that part of phase space between two surfaces of constant energy. Since the dynamical system is conservative, a point $A$ in this region will continue to remain in it throughout its history. If we start out with an initial uniform probability distribution on this region it will remain unchanged as time progresses. If we consider a phase function (or set of phase functions), that is, functions of $p_1, \ldots, p_n, q_1, \ldots, q_n$, as time progresses, a strictly stationary process is generated. It was in such a context that people tried to prove the equality of *space averages* and *time averages* and were led to ergodic theorems. We shall consider some ergodic theorems in section b of this chapter and in Chapter VII.

3. The previous examples dealt with continuous time parameter stationary processes. A simple example of a discrete time parameter stationary process is provided by a class of Markov chains. Let $\mathbf{P} = (p_{i,j}; i, j = 1, 2, \ldots)$ be a transition probability matrix with an invariant probability vector $\mathbf{p} = (p_i; i = 1, 2, \ldots)$, that is,

$$\mathbf{p}\mathbf{P} = \mathbf{p}. \tag{12}$$

As we have already seen, a Markov chain with stationary transition probability matrix $\mathbf{P}$ and invariant instantaneous probability vector $\mathbf{p}$ is an example of a stationary process (see Chapter III).

# b. The Ergodic Theorem and Stationary Processes

The ergodic hypothesis, "equality of space averages and time averages," was casually referred to in our discussion of Liouville's theorem. We shall now investigate this hypothesis in some detail in the context of a strictly stationary real-valued discrete parameter process. The restriction to a discrete time parameter is not essential and is used to avoid a tedious discussion of fine points.

Let $\{X_n\}, n = 0, \pm 1, \ldots,$ be a strictly stationary process with finite first moment, that is, $E|X_n| = E|X| < \infty$. The basic probability space can be thought of as the space of points $w = (\ldots, w_{-1}, w_0, w_1, \ldots)$ with $X_n(w) = w_n$. Thus each $w$ point represents a possible history of the system considered as it evolves from $t = -\infty$ to $t = +\infty$ with $X_n = w_n$ the location of the system at time $n$. We now introduce a transformation $T$ of the space $\Omega$ of $w$ points into itself. This transformation $T$ is naturally called the "shift" transformation since it takes the point $w$ with $n$-th coordinate $w_n$ into the point $w' = Tw$ with $n$-th coordinate $w'_n = (Tw)_n = w_{n+1}$. Let $T^{-1}$ be the transformation inverse to $T$ so that the point $w' = T^{-1}w$ has $n$-th coordinate $w'_n = (T^{-1}w)_n = w_{n-1}$.

Given any set $A$ of $w$ points let

$$TA = \{w | T^{-1}w \epsilon A\}. \tag{1}$$

By the stationarity assumption the event

$$A = \{w | X_{n_1}(w) \leq \lambda_1, \ldots, X_{n_k}(w) \leq \lambda_k\} \tag{2}$$

has the same probability as

$$\begin{aligned} T^{-1}A &= \{w | X_{n_1}(Tw) \leq \lambda_1, \ldots, X_{n_k}(Tw) \leq \lambda_k\} \\ &= \{w | X_{n_1+1}(w) \leq \lambda_1, \ldots, X_{n_k+1}(w) \leq \lambda_k\}. \end{aligned} \tag{3}$$

But then, by the theorem of Kolmogorov, $P(A) = P(T^{-1}A)$ for all sets $A$ in the Borel field generated by events of the form (2). Such a transformation $T$ is called a "measure-preserving" transformation for the probability measure $P(\cdot)$ since the probabilities of the events $A$ and $T^{-1}A$ are the same for all $A$. Thus, the shift transformation is a measure-preserving transformation for the probability measure of a strictly stationary process.

Ergodic theory is basically concerned with the limit behavior of time averages of the process

$$\frac{1}{n}\sum_{k=0}^{n-1} X_k(w) = \frac{1}{n}\sum_{k=0}^{n-1} X_0(T^k w) \tag{4}$$

or more generally, the limit behavior of time averages of a decent function of the process and its time shifts, that is,

$$\frac{1}{n}\sum_{k=0}^{n-1} f(T^k w). \tag{5}$$

By a decent function $f$, we mean one that is measurable with respect to the Borel field of measurable sets and is integrable so that $E|f(w)| < \infty$. Any function of the process can be written in the form $f(w)$ since the $w$ points are simply the countably dimensional vectors that are the possible sequences of $X_n$ values. A basic interest in ergodic theory is that of finding conditions under which the limiting time average

$$\lim_{n\to\infty} \frac{1}{n}\sum_{k=0}^{n-1} f(T^k w) \tag{6}$$

is equal to the "space average"

$$Ef(w) \tag{7}$$

for any decent function $f$ of the process. The expectation $Ef(w)$ is a space average since it is an average with respect to the probability measure of the process over all possible paths or histories $w$ of the system. On the other hand, the time average

$$\frac{1}{n}\sum_{k=0}^{n-1} f(T^k w) \tag{8}$$

is an average with respect to a specific history or sample point $w$. Of course, equality cannot be expected for all $w$; it is enough to require it for almost all $w$. A strictly stationary process $\{X_n\}$ is called an ergodic process if the limiting time average

$$\lim_{n\to\infty} \frac{1}{n}\sum_{k=0}^{n-1} f(T^k w) \tag{9}$$

*exists and is equal to the space average*

$$Ef(w) \tag{10}$$

*for almost all w and every integrable f.* Such ergodic processes are of considerable interest. For in the case of such a process by studying the time evolution of one possible sample path or history, the statistical structure of the whole ensemble of possible paths can be obtained due to the interchangeability of time and space averages.

The ergodic theorem that we shall obtain will give us necessary and sufficient conditions for the ergodicity of a process. It is unfortunately true that in many situations it may require almost as much work to verify these conditions as to prove ergodicity directly. However, there are probabilistic contexts in which these conditions are convenient. It is natural to introduce some of the relevant concepts in terms of the shift transformation $T$ and the probability measure $P$ of the process. *An event A of the Borel field is called an invariant event if $A = T^{-1}A$ with at most the exception of a set having P measure zero.* The collection of invariant events (or sets) is a subsigma-field of the Borel field of the process. *The process (or the corresponding measure P) is called "metrically transitive" if the field of invariant sets is the trivial field consisting of the whole space $\Omega$, the null set $\varphi$ and sets differing from these two by at most a set of probability or P measure zero.* It will later be seen that the metrically transitive processes are identical with the ergodic processes. Paralleling the notion of an invariant event, we introduce the notion of an invariant function. *A function f is called an invariant function if $f(w) = f(Tw)$ almost everywhere.* An equivalent way of stating this can be given in terms of invariant events. A function $f$ is invariant if the set $\{w|f(w) \leq x\}$ is invariant for every real number $x$. Notice that a process is metrically transitive if and only if every invariant function is equal to a constant almost everywhere. Thus in the case of metrically transitive processes, the only invariant functions are the trivial constant functions.

The ergodic theorem we prove in this section is called the "strong" or "individual" ergodic theorem because it is concerned with pointwise convergence. It was originally obtained by George Birkhoff. The proof given here is ingenious and is due to F. Riesz [62].

**Ergodic Theorem:** *Let $P( \cdot )$ be the measure on the space of sequences w generated by a strictly stationary process. Let $T$ be the shift transformation and $\mathcal{g}$ the Borel field of invariant w sets. If $f$ is a function with finite first moment*

$E|f(w)| < \infty$, *then*

$$\lim_{n \to \infty} \frac{1}{n} \sum_{j=0}^{n-1} f(T^j w) \tag{11}$$

*exists with probability one (P measure) and is equal to $E\{f(w)|\mathcal{s}\}$. If the stationary process is metrically transitive $E\{f(w)|\mathcal{s}\} = E\{f(w)\}$ so that time and space averages are equal.*

We first make a simple remark about a finite sequence $c_1, \ldots, c_n$ of real numbers. *Let A be the set of integers $m < n$ such that $c_m$ is exceeded by the maximum of those numbers in sequence following it*

$$c_m < \max_{j>m} c_j. \tag{12}$$

The set $A$ can then be decomposed into maximal disjoint blocks of successive integers. *If $\alpha, \beta$ are the first and last integers of such a block*

$$c_j < c_{\beta+1}, \ \alpha \leq j \leq \beta. \tag{13}$$

This is rather simple for since $\beta + 1 \notin A$, $c_{\beta+1} \geq \max_{j>\beta+1} c_j$, and therefore since $\beta \in A$

$$c_\beta < \max_{j>\beta} c_j = c_{\beta+1}. \tag{14}$$

If $\alpha < \beta$, the number $\beta - 1 \in A$ and hence $c_{\beta-1} < \max_{j>\beta-1} c_j = c_{\beta+1}$. For $\alpha < \beta - 1$ the argument continues as given above.

Consider now $f(w)$ any function with finite first moment $E|f(w)| < \infty$. Let $S_n = S_n(w)$ be the partial sum

$$S_n(w) = f(w) + f(Tw) + \cdots + f(T^{n-1}w). \tag{15}$$

Take $\beta$ to be any constant and $M$ any invariant set of the probability measure $P(\cdot)$ generated by the process. *As a basic step in the proof of the ergodic theorem, we shall first show that*

$$\int_{\left\{w \mid \text{L.u.b.}_{n \geq 1} \frac{S_n(w)}{n} > \beta\right\} M} f(w) \, dP \geq \beta P\left[\left\{\text{L.u.b.}_{n \geq 1} \frac{S_n(w)}{n} > \beta\right\} M\right]. \tag{16}$$

Here L.u.b. serves as a convenient abbreviation for least upper bound. Notice that it is enough to prove this inequality for the case $\beta = 0$, for one can reduce it to this case by replacing $f(w)$ by $f(w) - \beta$ and $S_n(w)/n$ by $S_n(w)/n - \beta$. Set

$$
\begin{aligned}
\Lambda &= \{w | \underset{n \geq 1}{\text{L.u.b.}}\ S_n(w) > 0\} \\
\Lambda_j &= \{w | \underset{1 \leq n \leq j}{\text{L.u.b.}}\ S_n(w) > 0\}.
\end{aligned}
\tag{17}
$$

The sequence of sets $\Lambda_j$ is an increasing sequence with $\Lambda$ the limit as $j \to \infty$. Consider the sequence $S_1(w), \ldots, S_n(w)$ and let $A = A(w)$ be the set of integers relative to this sequence discussed in the previous paragraph, that is,

$$
A(w) = \{m | m < n,\ S_m(w) < \max_{j > m} S_j(w)\}.
\tag{18}
$$

Further let $N_j$ be the set of $w$ points such that $j \epsilon A(w)$. The discussion of the preceding paragraph indicates that

$$
\sum_{j \epsilon A(w)} f(T^j w) \geq 0.
\tag{19}
$$

But this implies that

$$
\int_M \sum_{j \epsilon A(w)} f(T^j w)\ dP = \sum_{j=1}^{n-1} \int_{MN_j} f(T^j w)\ dP \geq 0.
\tag{20}
$$

The set

$$
N_j = \{w | \max_{j \leq k \leq n-1} [f(T^j w) + \cdots + f(T^k w)] > 0\} = T^{-j} \Lambda_{n-j}
\tag{21}
$$

and $T$ is measure-preserving. Therefore

$$
\begin{aligned}
0 \leq \sum_{j=1}^{n-1} \int_{MN_j} f(T^j w)\ dP &= \sum_{j=1}^{n-1} \int_{MT^{-j}\Lambda_{n-j}} f(T^j w)\ dP \\
&= \sum_{j=1}^{n-1} \int_{M\Lambda_{n-j}} f(w)\ dP = \sum_{j=1}^{n-1} \int_{M\Lambda_j} f(w)\ dP.
\end{aligned}
\tag{22}
$$

Since

$$
\lim_{j \to \infty} \int_{M\Lambda_j} f(w)\ dP = \int_{M\Lambda} f(w)\ dP
\tag{23}
$$

inequality (16) is obtained from (22) on dividing by $n$ and taking the limit as $n \to \infty$.

It is almost immediately clear that

$$
y_1(w) = \liminf_{n \to \infty} \frac{S_n(w)}{n}
\tag{24}
$$

and

$$
y_2(w) = \limsup_{n \to \infty} \frac{S_n(w)}{n}
\tag{25}
$$

are invariant functions. It will be enough to give the argument for $y_1(w)$. Now

$$y_1(w) = \lim_{n \to \infty} \inf \frac{f(w) + \cdots + f(T^{n-1}w)}{n}$$

$$= \lim_{n \to \infty} \inf \frac{f(Tw) + \cdots + f(T^n w)}{n} = y_1(Tw). \qquad (26)$$

We shall now show that $y_1(w) = y_2(w)$ almost everywhere so that $\lim_{n \to \infty} S_n(w)/n$ exists. Take any two real numbers $\alpha$, $\beta$, $\alpha < \beta$. Let $M_{\alpha,\beta}$ be the invariant set $\{w | y_1(w) < \alpha < \beta < y_2(w)\}$. Now

$$M_{\alpha,\beta} = M_{\alpha,\beta} \left\{ w \Big| \text{L.u.b.} \frac{S_n(w)}{n} > \beta \right\} \qquad (27)$$

so that by inequality (16)

$$\int_{M_{\alpha,\beta}} f(w) \, dP = \int_{M_{\alpha,\beta} \left\{ w | \text{L.u.b.} \frac{S_n(w)}{n} > \beta \right\}} f(w) \, dP \geq \beta P\{M_{\alpha,\beta}\}. \qquad (28)$$

The inequality

$$\int_{M_{\alpha,\beta}} f(w) \, dP \leq \alpha P\{M_{\alpha,\beta}\} \qquad (29)$$

is obtained by applying the same argument to $\{-f(T^j w)\}$ with $\alpha$, $\beta$ replaced by $-\beta$, $-\alpha$. The inequalities (28) and (29) cannot both be valid unless $P\{M_{\alpha,\beta}\} = 0$. Let us now take $\alpha$, $\beta$ any pair of rational numbers with $\alpha < \beta$. Then

$$P\{y_1(w) < y_2(w)\} = P\{ \bigcup_{\alpha < \beta} M_{\alpha,\beta}\} \leq \sum_{\alpha,\beta} P\{M_{\alpha,\beta}\} = 0 \qquad (30)$$

so that $y_1(w) = y_2(w) = y(w) = \lim_{n \to \infty} S_n(w)/n$ almost everywhere.

Since the random variables $f(T^j w)$ have a common distribution because of the stationarity of the measure $P$, the averages (see note 3 at the end of this chapter)

$$\frac{S_n(w)}{n} = \frac{f(w) + \cdots + f(T^{n-1}w)}{n} \qquad (31)$$

are uniformly integrable. But this implies that the limit $y(w) = \lim_{n \to \infty} S_n(w)/n$ is finite-valued with probability one and integrable. Take $M$ to be any invariant set. Then

$$\int_M f(w) \, dP = \frac{1}{n} \sum_0^{n-1} \int_{T^{-i}M} f(T^i w) \, dP = \frac{1}{n} \sum_0^{n-1} \int_M f(T^i w) \, dP$$

$$= \int_M \frac{S_n(w)}{n} \, dP. \qquad (32)$$

Because of the uniform integrability we can go to the limit and find that

$$\int_M f(w) \, dP = \int_M y(w) \, dP. \tag{33}$$

In fact, *the uniform integrability implies that $S_n(w)/n$ converges to $y(w)$ in the mean, that is,*

$$\int \left| \frac{S_n(w)}{n} - y(w) \right| dP \to 0 \tag{34}$$

*as $n \to \infty$.*

Since $y$ is measurable with respect to the sigma-field of invariant sets, it is clear that

$$y(w) = E(f(w)|\mathcal{I}) \tag{35}$$

and hence that $y(w) = Ef(w)$ almost everywhere if the measure is metrically transitive. On the other hand, if

$$\lim_{n \to \infty} \frac{1}{n} \sum_{j=0}^{n-1} f(T^j w) = Ef(w) \tag{36}$$

for every function $f$ with $E|f(w)| < \infty$, it is clear that the process must be metrically transitive. Thus, *metric transitivity is a necessary and sufficient condition for a stationary process to be ergodic.*

It will be convenient to obtain conditions for ergodicity of a stationary process in a different form. Let $I_A(w)$ be the set characteristic function of a measurable set $A$ (an event), that is,

$$I_A(w) = \begin{cases} 1 & \text{if } w \epsilon A \\ 0 & \text{otherwise.} \end{cases} \tag{37}$$

Let $A,B$ denote any two measurable sets. If the process is ergodic

$$\frac{1}{n} \sum_{j=0}^{n-1} I_A(T^j w) \to EI_A(w) = P(A) \tag{38}$$

as $n \to \infty$. But this implies that

$$\frac{1}{n} \sum_{j=0}^{n-1} I_A(T^j w) I_B(w) \to I_B(w) P(A) \tag{39}$$

as $n \to \infty$. On integrating the above, we find that

$$\frac{1}{n} \sum_{j=0}^{n-1} P(B \cap T^{-j}A) = \int \frac{1}{n} \sum_{j=0}^{n-1} I_A(T^j w) I_B(w) \, dP \qquad (40)$$
$$\to \int I_B(w) P(A) \, dP = P(B) P(A) \text{ as } n \to \infty.$$

Hence ergodicity implies that

$$\lim_{n \to \infty} \frac{1}{n} \sum_{j=0}^{n-1} P(B \cap T^{-j}A) = P(A)P(B) \qquad (41)$$

for any two measurable sets $A$, $B$. Now consider applying relation (41) to an invariant set $C$, that is, set $A$, $B = C$. Then

$$P(C) = \lim \frac{1}{n} \sum_{j=0}^{n-1} P(C \cap T^{-j}C) = P^2(C). \qquad (42)$$

Since one can only have $P(C) = 0$ or $P(C) = 1$, every invariant set is trivial and the process is ergodic by the ergodic theorem. *Condition* (41) *for any two measurable sets $A$, $B$ is* therefore *a necessary and sufficient condition for ergodicity.* Notice that (41) is implied by a stronger condition, namely

$$\lim_{j \to \infty} P(B \cap T^{-j}A) = P(B)P(A) \qquad (43)$$

for any two pairs of events $A$, $B$. This condition is commonly referred to as a *mixing condition* and is, essentially, a form of asymptotic independence.

As one might expect, *a stationary process of independent, identically distributed random variables is mixing.* We shall now prove this result. Let $A$, $B$ be any two events. Given any $\varepsilon > 0$, there are events $A_n$, $B_n$ measurable with respect to the Borel field $\mathfrak{B}_n$ generated by $X_{-n}(w)$, . . . , $X_n(w)$ ($n$ will of course depend on $\varepsilon$) such that the probabilities of the symmetric differences $A \ominus A_n$, $B \ominus B_n$ is less than $\varepsilon$, that is,

$$P(A \ominus A_n), P(B \ominus B_n) < \varepsilon. \qquad (44)$$

This follows since the union of the Borel fields $\mathfrak{B}_n$ is a field that generates the Borel field of the process. Now

$$|P(B \cap T^{-j}A) - P(B_n \cap T^{-j}A_n)| =$$
$$|P(B \cap T^{-j}A) - P(B_n \cap T^{-j}A) + P(B_n \cap T^{-j}A) - P(B_n \cap T^{-j}A_n)|$$
$$< 2\varepsilon. \qquad (45)$$

Since

$$P(B_n \cap T^{-j}A_n) = P(B_n)P(T^{-j}A_n) = P(B_n)P(A_n) \qquad (46)$$

for $j > 2n$, it follows that

$$|P(B \cap T^{-j}A) - P(B)P(A)| < 4\varepsilon \qquad (47)$$

for $j > 2n = 2n(\varepsilon)$. But this is valid for every $\varepsilon > 0$ and hence

$$\lim_{j \to \infty} P(B \cap T^{-j}A) = P(B)P(A). \qquad (48)$$

It is natural that a process of independent random variables ought to satisfy as strong a mixing condition as desired and thus be ergodic. A less trivial and somewhat more illuminating set of examples is provided by the class of stationary Markov chains. First consider the case of a chain that is not irreducible. Then there are at least two nonempty disjoint closed sets $C_1$, $C_2$ so that

$$0 < P(X_n(w)\epsilon C_i) < 1, \ i = 1, 2. \qquad (49)$$

Moreover, since $C_1$ is closed the event $A = \{w|X_0(w)\epsilon C_1\}$ is a nontrivial invariant set $(T^{-1}A = A)$. *A chain that is not irreducible is clearly not ergodic.* We shall now see that *the irreducible chains are the ergodic chains.* Ergodicity is implied by the validity of relation (41) for any pair of events $A$, $B$. However, by the argument of the preceding paragraph, it is enough to verify (41) for any pair of events $A_m$, $B_m$ measurable with respect to $\mathfrak{B}_m$ (the field generated by $X_{-m}(w), \ldots, X_m(w)$). Let $A_m$, $B_m$ be the events

$$\begin{aligned} A_m &= \{w|X_{-m}(w) = i_{-m}, \ldots, X_m(w) = i_m\} \\ B_m &= \{w|X_{-m}(w) = j_{-m}, \ldots, X_m(w) = j_m\}. \end{aligned} \qquad (50)$$

Now

$$P(B_m \cap T^{-k}A_m)$$
$$= p_{i_{-m}}p_{i_{-m},i_{-m+1}} \cdots p_{i_{m-1},i_m}p_{i_m,j_{-m}}^{(k-2m)}p_{j_{-m},j_{-m+1}} \cdots p_{j_{m-1},j_m} \qquad (51)$$

for $k$ sufficiently large. By Problem 15 of Chapter III, if the chain is irreducible

$$\lim_{n \to \infty} \frac{1}{n} \sum_{k=0}^{n-1} p_{i_m,j_{-m}}^{(k)} = p_{j_{-m}}. \qquad (52)$$

But this implies that

$$\lim_{n \to \infty} \frac{1}{n} \sum_{k=0}^{n-1} P(B_m \cap T^{-k}A_m) \qquad (53)$$

$$= p_{i_{-m}}p_{i_{-m},i_{-m+1}} \cdots p_{i_{m-1},i_m}p_{j_{-m}}p_{j_{-m},j_{-m+1}} \cdots p_{j_{m-1},j_m} = P(A_m)P(B_m)$$

so that the chain is ergodic. *If the chain is irreducible with persistent states,* condition (52) can be replaced by the stronger condition

$$\lim_{n \to \infty} p^{(n)}_{im, j-m} = p_{j-m},\tag{54}$$

which implies that *the chain is mixing by a similar argument.*

Let us now return to the case of a stationary process of independent random variables and make an almost obvious but nonetheless worthwhile remark. Any Borel function $f(w)$ on the sample space of the process induces a derived process, namely,

$$Y_n(w) = f(T^n w).\tag{55}$$

Since the original process $X_n(w)$ is mixing, it follows that *any such derived process $Y_n(w)$ is mixing.* In particular, any process of the form

$$Y_n = \sum_{j=-\infty}^{\infty} a_j X_{n-j}, \quad \sum_j a_j^2 < \infty,\tag{56}$$

with the $X_n$'s independent, identically distributed with mean zero and variance one, is mixing.

There are not many processes that have been characterized in terms of ergodic or mixing properties. A precise characterization of the normal processes that are ergodic or mixing has been given in a paper of Maruyama [55].

## c. Convergence of Conditional Probabilities

At this point we shall prove a theorem on the convergence of sequences of conditional probabilities. Even though there is considerable interest in this result for its own sake, we shall be basically interested in applying it to obtain MacMillan's theorem, a result of some importance in information theory, in the following section. It should be noted that this convergence theorem can be regarded as a special case of a Martingale convergence theorem (see Doob [12]).

Suppose there is an increasing sequence of subsigma-fields $\mathcal{C}_n(\mathcal{C}_n \subset \mathcal{C}_{n+1})$ of $\mathcal{B}$ with $\mathcal{C}$ the subsigma-field of $\mathcal{B}$ that $\cup \, \mathcal{C}_n$ generates. Let $A$ be an event. *We shall show that*

$$P(A|\mathcal{C}_n) \to P(A|\mathcal{C})\tag{1}$$

*almost everywhere as $n \to \infty$.* In section d this result will be applied in a context where $\mathcal{C}_n$ is the Borel field generated by a family of random variables $X_{-1}(w)$, . . . , $X_{-n}(w)$ and $\mathcal{C}$ that generated by the infinite family $X_{-1}(w)$, $X_{-2}(w)$, . . . . Notice that (1) can be rewritten

$$E(c_A(w)|\mathcal{C}_n) \rightarrow E(c_A(w)|\mathcal{C}) \tag{2}$$

where $c_A$ is the set-characteristic function of the event $A$. This suggests the apparently more general result

$$E(f(w)|\mathcal{C}_n) \rightarrow E(f(w)|\mathcal{C}) \tag{3}$$

with $f$ a bounded measurable function (with respect to $\mathcal{B}$). Notice that (3) readily follows if we are able to show that

$$E(g(w)|\mathcal{C}_n) \rightarrow g(w) \tag{4}$$

almost everywhere for a bounded function $g$ measurable with respect to the subsigma-field $\mathcal{C}$. For $g(w) = E(f(w)|\mathcal{C})$ is measurable with respect to $\mathcal{C}$ and hence by a basic property of conditional expectations (see equation (IV.c.12))

$$E(f(w)|\mathcal{C}_n) = E(E(f(w)|\mathcal{C})|\mathcal{C}_n) = E(g(w)|\mathcal{C}_n)$$
$$\rightarrow g(w) = E(f(w)|\mathcal{C}) \tag{5}$$

as $n \rightarrow \infty$.

Further, it is enough to prove (4) for simple functions (linear combination of a finite number of set characteristic functions of events) measurable with respect to $\mathcal{C}$ and therefore sufficient to verify it for a $\mathcal{C}$ measurable characteristic function. For a bounded $\mathcal{C}$ measurable function $g$ is the limit almost everywhere of a uniformly convergent sequence of simple functions $g_k$ measurable $\mathcal{C}$ (such a sequence is given by $g_k(w) = [kg(w)]/k$ where $[x]$ denotes the largest integer less than $x$) and

$$|E(g|\mathcal{C}_n) - g| \leq E(|g - g_k| \,|\mathcal{C}_n) + |E(g_k|\mathcal{C}_n) - g_k| + |g_k - g|. \tag{6}$$

The first and last term on the right will be small if $k$ is large. The second term is small for large $n$.

Since it is enough to verify (2) for $g$ the characteristic function $c_A(w)$ of a set $A$ measurable $\mathcal{C}$, let us establish this. Take $\varepsilon$, $\delta$ as two small positive numbers less than one. Since $\mathcal{C}$ is generated by the fields $\mathcal{C}_n$, there is a set $B$ in some field $\mathcal{C}_k$ with $k$ sufficiently large such that $P(A \ominus B) < \varepsilon \delta/2$. Let

$$D_n = \{w|1 - P(A|\mathcal{C}_n)(w) \geq \varepsilon\}. \tag{7}$$

Set $F_k^{(1)} = B \cap D_k$, $\quad F_k^{(2)} = B \cap D_{k+1} - F_k^{(1)}$, $\quad F_k^{(3)} = B \cap D_{k+2} -$
$(F_k^{(1)} \cup F_k^{(2)})$, . . . . Notice that $F_k^{(j)}$ is in $\mathcal{C}_{k+j-1}$. Now

$$P(F_k^{(j)} - A) = \int_{F_k^{(j)}} [1 - P(A|\mathcal{C}_{k+j-1})]\, dP \geq P(F_k^{(j)})\, \varepsilon \qquad (8)$$

so that if $F_k = \bigcup_j F_k^{(j)}$, then

$$P(F_k - A) \geq P(F_k)\, \varepsilon. \qquad (9)$$

This implies that $P(B - A) \geq P(F_k)\, \varepsilon$. But $P(F_k) < \delta/2$ since $P(B - A)$
$\leq P(B \ominus A) < \varepsilon\, \delta/2$.

If $w \epsilon B - F_k$, then $P(A|\mathcal{C}_n) \geq 1 - \varepsilon$ when $n \geq k$ for $F_k = B \cap$
$\{w | 1 - \inf\limits_{n \geq k} P(A|\mathcal{C}_n) \geq \varepsilon\}$ where inf is an abbreviation for infimum.
This implies that $P(A|\mathcal{C}_n) \geq 1 - \varepsilon$ throughout $A$ except for a set of
probability at most $\delta$ since $P(A \ominus B) + P(F_k) \leq \varepsilon\, \delta/2 + \delta/2 < \delta$.
Since $\varepsilon$, $\delta$ are arbitrary $P(A|\mathcal{C}_n)$ converges to 1 almost everywhere in $A$.
The same argument applied to $\bar{A}$ indicates that $P(A|\mathcal{C}_n) \to 0$ almost
everywhere in $\bar{A}$. Thus $P(A|\mathcal{C}_n) \to c_A(w)$ almost everywhere.

## d. MacMillan's Theorem

In recent years there has been a considerable interest in communication problems. In particular, work of Shannon [73] and others has
led to the growth of a field of interest commonly referred to as "information theory." Excellent discussions of the basic problems in this
field are to be found in the monographs of Khinchin [42] and A. Feinstein [14]. In such problems, there is a message to be transmitted over
a communication channel. The message must be encoded in a form
that the channel can accept. After transmission, the message received
must be decoded so as to reproduce the information transmitted. Since
there is noise corrupting and distorting the transmission of messages
over the channel, a basic question that arises is the encoding of the
message so as to guarantee the most efficient transmission of information over the channel. This will depend on the statistical character
of the message to be transmitted and the channel. We shall not study
the full communication problem, only certain aspects of the message
to be transmitted.

Suppose the message to be transmitted is in an alphabet of $s$
letters. For convenience, let us denote these letters by the symbols
$a_1, \ldots, a_s$. Assume that the message to be transmitted can be rea-

sonably regarded as the realization of a stationary stochastic process $X_n(w)$, $n = 0, \pm 1, \pm 2, \ldots$ . Then $X_n(w)$ is the symbol to be trans- mitted at time $n$. The probability space of the process $X_n(w)$ has a nonenumerable number of elementary events and the concept of entropy has only been introduced for probability spaces with a finite (or countable) number of elementary events. It will be advantageous to extend the entropy concept to the probability spaces of such processes $X_n(w)$.

Consider the random variables $X_{-n+1}(w)$, $\ldots$ , $X_0(w)$. There are $s^n$ possible corresponding distinct sequences of symbols and hence the field $\mathcal{C}_n$ of these random variables can be regarded as a finite field with $s^n$ elementary events. Let $T$ be the shift operator introduced in section b. Then $T\mathcal{C}_n$ is the field of $X_{-n}(w)$, $\ldots$ , $X_{-1}(w)$. Let $H_n = H(\mathcal{C}_n)$ be the entropy of $\mathcal{C}_n$. *If* $\lim\limits_{n \to \infty} H_n/n = H$ *exists, it would be*

natural *to call this limit the entropy of the process* $X_n(w)$ (it would be more accurate to call this the average entropy per symbol). *This limit will be shown to exist.* Now

$$H(\mathcal{C}_{n+m}) = H(\mathcal{C}_n \vee T^n \mathcal{C}_m)$$
$$= H(\mathcal{C}_n) + H_{\mathcal{C}_n}(T^n \mathcal{C}_m). \tag{1}$$

Further

$$H_{\mathcal{C}_n}(T^n \mathcal{C}_m) \leq H(T^n \mathcal{C}_m) = H(\mathcal{C}_m) \tag{2}$$

(see problem 19 of Chapter II) from which it follows that

$$H_n = H(\mathcal{C}_n) \leq H_{n+m} = H(\mathcal{C}_{n+m}) \leq H_n + H_m = H(\mathcal{C}_n) + H(\mathcal{C}_m). \tag{3}$$

From this it follows that $H_n \leq nH_1$ and hence that

$$H = \lim_{n \to \infty} \inf H_n/n \tag{4}$$

is finite. Consider any $\varepsilon > 0$. Let $m$ be such that

$$\frac{H_m}{m} < H + \varepsilon. \tag{5}$$

Given any $n > m$, there is an integer $k > 1$ such that

$$(k - 1)m \leq n < km. \tag{6}$$

But then inequality (3) implies that

$$\frac{H_n}{n} \leq \frac{H_{km}}{(k-1)m} \leq \frac{k}{k-1} \frac{H_m}{m} < \frac{k}{k-1} (H + \varepsilon) \tag{7}$$

and therefore, for sufficiently large $n$,

$$H - \varepsilon < \frac{H_n}{n} < \frac{k}{k-1}(H + \varepsilon) < H + 2\varepsilon. \tag{8}$$

Since $\varepsilon$ is arbitrary

$$H = \liminf_{n \to \infty} \frac{H_n}{n} = \lim_{n \to \infty} \frac{H_n}{n}. \tag{9}$$

Now let

$$p_n(\alpha_0, \ldots, \alpha_{n-1}) = P[X_0(w) = \alpha_0, \ldots, X_{n-1}(w) = \alpha_{n-1}]$$
$$p(\alpha_{n-1}|\alpha_{n-2}, \ldots, \alpha_0)$$
$$= P[X_{n-1}(w) = \alpha_{n-1}|X_{n-2} = \alpha_{n-2}, \ldots, X_0(w) = \alpha_0] \tag{10}$$

where $\alpha_0, \ldots, \alpha_{n-1}$ each assume one of the values $a_1, \ldots, a_s$.
*MacMillan's theorem states that the random quantity*

$$-\frac{1}{n} \log p_n(X_0(w), \ldots, X_{n-1}(w)) = f_n(w) \tag{11}$$

*converges to $H$ in mean as $n \to \infty$ if $X_n(w)$ is an ergodic process.* This result
is essentially obtained by an application of the ergodic theorem.
    We have

$$-\log p_n(X_0(w), \ldots, X_{n-1}(w))$$
$$= -\sum_{k=0}^{n-1} \log p(X_k(w)|X_{k-1}(w), \ldots, X_0(w)) \tag{12}$$
$$= \sum_{k=0}^{n-1} g_k(T^k w)$$

where

$$g_k(w) = -\log p(X_0(w)|X_{-1}(w), \ldots, X_{-k}(w)). \tag{13}$$

By section c,

$$-\log P(X_0(w) = \alpha|X_{-1}(w), \ldots, X_{-k}(w)) = g_k(w;\alpha) \tag{14}$$

converges almost everywhere as $k \to \infty$. But then

$$g_k(w;\alpha) - g_{k'}(w;\alpha) \to 0 \tag{15}$$

almost everywhere as $k,k' \to \infty$. However,

$$|g_k(w) - g_{k'}(w)| \leq \sum_{\alpha=a_1}^{a_s} |g_k(w;\alpha) - g_{k'}(w;\alpha)| \to 0 \tag{16}$$

as $k,k' \to \infty$ and therefore $g_k(w)$ converges to a limit $g(w)$ almost
everywhere as $k \to \infty$. It is conceivable that $g(w)$ might be infinite
on a set of positive probability. As we shall see this is not possible since

the sequence $g_k(w)$ will be shown to be uniformly integrable. Notice that $g_k(w)$ is measurable with respect to $\mathcal{Q}_{k+1}$. Let $B_k$ be an elementary event in $T\mathcal{Q}_k$ and $A_1$ an elementary event in $\mathcal{Q}_1$. An elementary event of $\mathcal{Q}_{k+1}$ is then of the form $B_k A_1$. Since $g_k(w)$ is measurable with respect to $\mathcal{Q}_{k+1}$, for $w \epsilon B_k A_1$,

$$g_k(w) = -\log \frac{P(B_k A_1)}{P(B_k)}. \tag{17}$$

Let $S_{k,j} = \{w | j \leq g_k(w) < j + 1\}$. If $B_k A_1$ is an elementary event of $\mathcal{Q}_{k+1}$ that is a subset of $S_{k,j}$ then

$$-\log \frac{P(B_k A_1)}{P(B_k)} \geq j \tag{18}$$

and hence $P(B_k A_1) \leq e^{-j} P(B_k)$. Then

$$\int_{S_{k,j}} g_k(w)\, dP = \sum_{B_k} \int_{B_k S_{k,j}} g_k(w)\, dP$$
$$\leq \sum_{B_k, A_1} (j + 1) P(B_k A_1 S_{k,j}) \leq s(j + 1) e^{-j} \tag{19}$$

where $s$ is the number of values that can be assumed by $X_n(w)$. Inequality (19) implies that the sequence $g_k(w)$ is a uniformly integrable sequence of random variables. Thus $g(w) = \lim g_k(w)$ is integrable and

$$\int |g_k(w) - g(w)|\, dP \to 0 \tag{20}$$

as $k \to \infty$. By the ergodic theorem

$$\frac{1}{n} \sum_{k=0}^{n-1} g(T^k w) \tag{21}$$

approaches an invariant function $h(w)$ in mean as $n \to \infty$. Further, if the process $X_n(w)$ is ergodic, $h(w)$ is the constant

$$h(w) = Eg(w) = \lim_{k \to \infty} Eg_k(w) = H. \tag{22}$$

Now

$$\int |f_n(w) - h(w)|\, dP = \int \left| \frac{1}{n} \sum_{k=0}^{n-1} g_k(T^k w) - h(w) \right| dP$$
$$\leq \int \left| \frac{1}{n} \sum_{k=0}^{n-1} [g_k(T^k w) - g(T^k w)] \right| dP + \int \left| \frac{1}{n} \sum_{k=0}^{n-1} g(T^k w) - h(w) \right| dP$$
$$\leq \frac{1}{n} \sum_{k=0}^{n-1} \int |g_k(w) - g(w)|\, dP + \int \left| \frac{1}{n} \sum_{k=0}^{n-1} g(T^k w) - h(w) \right| dP \to 0 \tag{23}$$

as $n \to \infty$ by the stationarity of the measure $P$ (generated by the $X_n(w)$ process) and the ergodic theorem. The proof of MacMillan's theorem is complete.

MacMillan's theorem can be given an interesting interpretation. Consider messages of length $n$. There are $s^n$ such possible messages. MacMillan's theorem states that with probability almost one the probability of such a possible message is to the first order

$$e^{-nH}. \tag{24}$$

There are roughly $e^{nH}$ such sequences. However, $H \leq \log s$ and thus the number of messages of appreciable probability is $e^{nH}$ which is usually much smaller than $s^n = e^{n \log s}$. In encoding, one therefore need essentially consider only the $e^{nH}$ messages of large probability.

# e. Problems

1. Consider the integers 1, 2, . . . , $n$ with the uniform distribution on them. Let the function $f$ map the integers 1, 2, . . . , $n$ onto themselves. Show that the process generated by $f$ is strictly stationary. Under what circumstances is it ergodic? Can it be mixing?

2. Let $X$ be a random variable uniformly distributed on $[0,1]$. Consider the process $X_n = X + n\alpha \bmod 1$ where $\alpha$ is an irrational number. Show that the process $X_n$, $n = 0, \pm 1, \ldots$ is strictly stationary. Is it ergodic?

3. Let $X$ be a given random variable. Consider the process $X_n = X$, $n = 0, \pm 1, \ldots$ . Is this process ergodic? If not, indicate the sigma-field of invariant sets.

4. Consider the unit square $0 \leq x, y \leq 1$ with uniform measure on it. Let $T(x,y) = (2x, \frac{1}{2}y)$ if $0 \leq x < \frac{1}{2}$ and $T(x,y) = (2x, \frac{1}{2}(y + 1))$ if $\frac{1}{2} \leq x < 1$. These equations are to be taken mod 1. Show that the process generated by the transformation is strictly stationary. Is it ergodic?

5. We say $X_n$ is $m$-step dependent if blocks of random variables are independent when separated by $m$ indices. Let $X_n$ be an $m$-step dependent strictly stationary process. When is such a process ergodic? When is it mixing? Construct examples of $m$-step dependent processes for $m = 1, 2, \ldots$ .

**6.** Let $X_n$ be a strictly stationary process that can assume only the values 0 or 1 at any fixed time. Let the probability

$$P[X_{n_1} = i_1, X_{n_2} = i_2, \ldots, X_{n_k} = i_k]$$

$$= \int_0^1 p^{\sum\limits_{\alpha=1}^{k} i_\alpha} (1 - p)^{k - \sum\limits_{\alpha=1}^{k} i_\alpha} dF(p)$$

where $F(p)$ is a distribution function with all its mass on [0,1]. What are the invariant sets of this process? Under what conditions is it ergodic?

**7.** Compute the entropy per unit time of a stationary Markov chain.

# Notes

1. An extensive discussion of the background in statistical mechanics in which the ergodic problems first arose can be found in Khinchin's book [41]. M. Kac has much material on the current status of allied problems in statistical mechanics in his recent volume [37].

2. Ergodic theory has grown into a field of considerable breadth after G. Birkhoff and J. von Neumann derived their ergodic theorems. It would be hopeless to try to give any extensive bibliography of the work in this area. It is worthwhile, however, referring to the books of E. Hopf [35] and P. Halmos [30] on ergodic theory. Hopf's monograph is excellent in its treatment of the early work on these problems. Halmos's essay is recent and discusses interesting open questions in this area. The two books are an excellent introduction to ergodic theory. A later book of S. R. Foguel [A5] is concerned with ergodic theory of a Markov process and deals with results and ideas that we haven't touched on.

3. The uniform integrability of the sequence $S_n/n$ used on page 108 can be readily seen in the following way. The random variables $f(T^j w)$ are uniformly integrable since their marginal distribution functions are identical. Thus, given any $\epsilon > 0$, there is a $\delta(\epsilon) > 0$ such that if $P(M_n) < \delta(\epsilon)$

$$\int_{M_n} |f(T^j w)| \, dP < \epsilon, \qquad j = 0, \ldots, n - 1.$$

But then

$$\int_{M_n} \left| \frac{S_n}{n} \right| dP \leq \frac{1}{n} \sum_{j=0}^{n-1} \int_{M_n} \left| \frac{f(T^j w)}{n} \right| dP < \epsilon.$$

# VI

## MARKOV PROCESSES

### a. Definition

We have already discussed a special but exceedingly rich and interesting class of Markov processes in Chapter III, the Markov chains. In going from Markov chains to Markov processes one allows possibly for a much more general state space. Continuous time parameter processes will be of special interest in this chapter.

The object of this section is to characterize the class of processes for which past and future are independent given precise knowledge of the present. It is already clear from the preceding statement that the indexing parameter of the random variables of the process must be a time-like parameter. Let $T$ be the parameter set. The most common choices of $T$ are the set of real numbers or the set of lattice points $kh$, $k = 0, \pm 1, \ldots, h > 0$, or any subset of these. Now consider the question of specifying the joint probability structure of the random variables $\{X_t(w), t \epsilon T\}$. This will be done by specifying the joint probability structure of any finite number of them in a consistent way. The theorem of Kolmogorov [44] then assures us that there is a stochastic process with this probability structure. Actually we shall not restrict ourselves to real-valued random variables. The random variables $X_t(w)$ can take as values complex numbers, points in Euclidean $k$-dimensional space, or even points in an abstract space. However, even if we do allow this generality, it is clear we have to say something about the range of the random variables, that is, the state space of the process. Let $\Omega_X$ be the state space of the random variables $X_t$. Take $\mathfrak{F}_X$ as the sigma-field of measurable subsets of $\Omega_X$. Further, we shall assume that $\mathfrak{F}_X$ includes all sets consisting of a single point of $\Omega_X$. For convenience, assume that there is a first point to the set $T$. The probability structure will be specified in terms of an initial probability measure and a transition probability function describing how transitions take place from one time to another.

Let $P(t_0, A)$ be a probability measure on the sets $A$ of $\mathfrak{F}_X$. This, as we shall see, is the probability distribution at the initial time $t_0$. Further let the transition probability function $P(\tau,x;t,A)$, $t_0 \leq \tau < t$, $x \epsilon \Omega_X$, $A \epsilon \mathfrak{F}_X$, be a function with the following properties:

   (i) $P(\tau,x;t,A)$ is a probability measure in $A \epsilon \mathfrak{F}_X$ for fixed $\tau$, $x$, $t$;
   (ii) $P(\tau,x;t,A)$ is measurable in $x$ with respect to $\mathfrak{F}_X$ for fixed $\tau$, $t$, $A$;
   (iii) $P(\tau,x;t,A)$ satisfies the integral equation (commonly called the Chapman-Kolmogorov equation)

$$P(\tau,x;t,A) = \int_{\Omega_x} P(\tau,x;t',dy) P(t',y;t,A) \tag{1}$$

for any $t'$ with $\tau < t' < t$. We are now in a position to specify the joint probability structure of the random variables $\{X_t, t \epsilon T\}$. Of course, $X_t$ is to specify the location of an observed system subject to random disturbances at time $t$. The probability that the system is in set $A_0$ at time $t_0$ and in set $A_j$ at time $t_j$, $t_0 < t_1 < \cdots < t_n$, is defined to be

$$P[X_{t_0} \epsilon A_0, \ldots, X_{t_n} \epsilon A_n] = \int_{A_0} \cdots \int_{A_{n-1}} P(t_0,dy_0) P(t_0,y_0;t_1,dy_1) \tag{2}$$
$$\cdots P(t_{n-2},y_{n-2};t_{n-1},dy_{n-1}) P(t_{n-1},y_{n-1};t_n,A_n).$$

A little reflection indicates that (2) coupled with the Chapman-Kolmogorov equation ensure that the probability structure is given consistently. The probability distribution $P(t,A)$ at time $t > t_0$ is then

$$P(t,A) = P[X_t \epsilon A] = \int_{\Omega_x} P(t_0,dy_0) P(t_0,y_0;t,A). \tag{3}$$

The family of sets $\{w | X_{t_0}(w) \epsilon A_0, \ldots, X_{t_n}(w) \epsilon A_n\}$ generates a field of sets in the space of sample functions $w = w(t)$ (see section d of Chapter IV) with values of the sample functions points in the space $\Omega_X$. Now since $P$ as given by (2) is countably additive on this field, it can by Caratheodory's theorem be extended to a probability measure on the sigma-field generated by this field.

In the case of Markov chains (see Chapter III) time is discrete and the state space is countable. Because of this, the transition probability functions can be represented in matrix form and integration is replaced by summation in the formulas given above.

It follows readily from (2) that the transition probability function $P(\tau,x;t,A)$ is the conditional probability $P[X_t(w) \epsilon A | X_\tau(w) = x]$. Suppose we take sets $A$, $B$ of the form

$$A = \{w | X_{t_1}(w) \epsilon A_1, \ldots, X_{t_k}(w) \epsilon A_k\}$$
$$B = \{w | X_{t_{k+1}}(w) \epsilon A_{k+1}, \ldots, X_{t_n}(w) \epsilon A_n\}, \tag{4}$$

$t_0 \le t_1 < \cdots < t_k < t < t_{k+1} < \cdots < t_n$, (as in section a of Chapter III) in the past and future relative to $t$. Then (2) implies that

$$P[AB|X_t(w) = x] = P[A|X_t(w) = x]P[B|X_t(w) = x] \qquad (5)$$

so that given the present the past and future are independent. There is a second way of rephrasing the Markovian property that follows from (2). Let $B$ be an event in the future relative to $t$. It then follows from (2) that

$$P[B|X_t(w) = x, X_{t_k}(w) = x_k, \ldots, X_{t_1}(w) = x_1]$$
$$= P[B|X_t(w) = x], \qquad (6)$$

that is, all higher order conditional probabilities reduce to first-order conditional probabilities. Of course, this is not surprising since (2) states that the full probability structure is determined by the initial probability distribution and the first-order conditional probabilities in the case of a Markov process. In the case of a non-Markovian process, the higher order conditional probabilities do not reduce to first-order conditional probabilities and hence they are needed for a full specification of the probability structure of the process.

If the transition probability function depends on $\tau$, $t$ only through the time difference $t - \tau$ so that

$$P(\tau, x; t, A) = P(t - \tau, x, A), \qquad (7)$$

the process $\{X_t\}$ is said to have a *stationary transition mechanism*. The Brownian motion and Poisson processes introduced in section d of Chapter IV are examples of Markov processes with stationary transition mechanism. In the case of the Brownian motion process, the state space is the set of real numbers and the transition probability function is given by

$$P(\tau, x; t, A) = P[X_t(w)\epsilon A|X_\tau(w) = x]$$
$$= P(t - \tau, x, A) = \int_A \frac{1}{\sqrt{2\pi(t - \tau)}} \exp\left(-\frac{(y - x)^2}{2(t - \tau)}\right) dy. \qquad (8)$$

The state space of the Poisson process is the set of non-negative integers. The transition probability function is given by

$$P(t - \tau, j, A) = \sum_{\substack{k\epsilon A \\ k \ge j}} \frac{[\lambda(t - \tau)]^{k-j}}{(k - j)!} e^{-\lambda(t-\tau)} \qquad (9)$$

where $k$ and $j$ are non-negative integers. Notice that even though the two specific Markov processes referred to above have stationary transition mechanism, they are not stationary processes. In fact, a Markov process with stationary transition mechanism will not be stationary unless its probability distribution at time $t$ $P(t,A)$ is independent of $t$, that is,

$$P(t,A) = \int_{\Omega x} P(t_0,dx)P(t - t_0, x, A)$$
$$= P(t_0,A) = P(A). \tag{10}$$

Thus, the initial distribution $P(t_0,A) = P(A)$ must reproduce itself through time. However, we have already seen in section a of Chapter III that there are stationary transition probability functions for which there is no self-reproducing initial probability distribution.

If the Markov process has stationary transition mechanism and time is discrete ($t = 0, \pm 1, \ldots$), all higher order transition probabilities $P(t,x,A)$ can be given recursively in terms of the first-order or one-step transition probability $P(1,x,A)$ as follows for $t = 0, 1, \ldots$

$$P(2,x,A) = \int_{\Omega x} P(1,x,dy)P(1,y,A)$$

$$\cdots$$

$$P(t + 1, x, A) = \int_{\Omega x} P(t,x,dy)P(1,y,A) \tag{11}$$
$$= \int_{\Omega x} P(1,x,dy)P(t,y,A).$$

It is very easy to give examples of non-Markovian processes. For simplicity consider discrete time $t = 0, \pm 1, \ldots$ Let $X_t$ be a normal process satisfying the difference equation

$$\sum_{k=0}^{m} a_k X_{t-k} = \xi_t, \qquad a_0, a_m \neq 0, \tag{12}$$

where the $\xi_t$ are independent normal random variables with mean zero and $\xi_t$ is independent of $X_{t-1}, X_{t-2}, \ldots$ The process $X_t$ is clearly Markovian if $m = 1$ but not if $m > 1$. For the conditional expected value $E(X_t|X_{t-1}, \ldots, X_{t-m}) = -a_0^{-1} \sum_{k=1}^{m} a_k X_{t-k}$ and this would depend only on $X_{t-1}$ if the process were Markovian. However, even if $m > 1$ we can introduce a related process with the same amount of information that is Markovian. Let $Y_t$ be the $m$-vector valued process

$$\mathbf{Y}_t = (X_t, \ldots, X_{t-m+1}). \tag{13}$$

This process is clearly Markovian. Thus, a process $\mathbf{Y}_t$ with a simpler structure has been obtained, a Markovian structure, at the cost of a more complicated state space. This can generally be done in an almost vacuous manner for any process $X_t$, $t = 0, \pm 1, \ldots$ Simply let $\mathbf{Y}_t$, $t = 0, \pm 1, \ldots$, be the process with

$$\mathbf{Y}_t = (X_t, X_{t-1}, \ldots). \tag{14}$$

The process $\mathbf{Y}_t$ is a Markov process with state space consisting of points with a countably infinite number of components. This trivial way of imbedding a general process in a Markov process usually does not lead to anything interesting. However, there is a variety of problems dealing with stochastic processes where the solution basically depends on imbedding the process cleverly in a Markov process.

There have in recent years been attempts to extend a Markov like notion to processes with a multidimensional time parameter. We briefly describe such a definition and refer to a paper of L. Pitt [A12] for a detailed exposition and development as well as a bibliography of related papers. Let $X_{t_1, t_2}$ be a process on the plane. Consider any domain with a smooth boundary. Assume any event determined by observations in the domain and any event determined by observations outside the domain are conditionally independent given data on the boundary. If this is true for every such domain, the process is called Markovian.

# b. Jump Processes with Continuous Time

A systematic study of continuous time parameter Markov processes was given by Kolmogorov in his basic paper [43]. Among these were Markov processes with a countable state space. The characteristic feature of these processes is that they are jump processes, that is, in a small time, the system described by the process is almost sure to remain in the state originally occupied and it will jump to another state only with small probability. The scope of such jump processes was enlarged by Feller in his papers [15] and [16]. We shall, in fact, generally follow the approach given in Feller's paper [16]. Our first object is to indicate more precisely what is meant by a jump process. Our discussion will be carried out in the context of a general state space $\Omega_X$. The reader may find it helpful to keep the case of a denumerable state space in mind in which case all integrals over the state space are to be replaced by sums over the state space. Functions $p(t,x) \geq 0$ and $\Pi(t,x,A)$, with $t$

real, $x\epsilon\Omega_X$ and $A\epsilon\mathfrak{F}_X$ satisfying the following conditions are assumed given:

(i) For fixed $x$, $p(t,x)$ is a continuous function of $t$ and for fixed $t$ it is measurable with respect to $\mathfrak{F}_X$. We shall assume that $p(t,x)$ is bounded in $t$, $x$ jointly so that one is led to a fairly simple theory. Otherwise the study of jump processes becomes much more complicated (see section e of this chapter for an example in which such complications arise).

(ii) For fixed $x$, $A$, $\Pi(t,x,A)$ is continuous in $t$; for fixed $t$, $x$, it is a probability measure on $\mathfrak{F}_X$; for fixed $t$, $A$, $\Pi(t,x,A)$ is measurable with respect to $\mathfrak{F}_X$. The jump character of the process is determined by specifying the transition behavior of the process in a small time interval. The process is a jump process if the transition probability

$$P(\tau,x;t,A) = \{1 - p(t,x)(t - \tau)\}\delta(x,A)$$
$$+ p(t,x)(t - \tau)\Pi(t,x,A) + o(t - \tau) \qquad (1)$$

for small $t - \tau > 0$, where

$$\delta(x,A) = \begin{cases} 1 \text{ if } x\epsilon A \\ 0 \text{ otherwise.} \end{cases} \qquad (2)$$

Notice that the probability of a change of state in time $t - \tau$ is of order $t - \tau$. A Poisson process is perhaps the simplest and most typical nontrivial example of such a discontinuous process. In the case of the Poisson process, the state space consists of the non-negative integers. There $p(t,x) \equiv \lambda$ and $\Pi(t,x,A) = \delta(x + 1, A)$. Thus, if there is a change of state in a small time, it is to the first order in $t - \tau$ from $x$ to $x + 1$.

The transition probability function $P(\tau,x;t,A)$ of a Markov process satisfying the assumptions of this section can be shown to be a solution of two integrodifferential equations

$$\frac{\partial P(\tau,x;t,A)}{\partial \tau} = p(\tau,x)\left\{P(\tau,x;t,A) - \int P(\tau,y;t,A)\Pi(\tau,x,dy)\right\} \qquad (3)$$

and

$$\frac{\partial P(\tau,x;t,A)}{\partial t} = -\int_A p(t,y)P(\tau,x;t,dy) + \int p(t,y)\Pi(t,y,A)P(\tau,x;t,dy), \qquad (4)$$

commonly called the backward and forward equations respectively. The derivation of the backward equation will be given in some detail. That of the forward equation will be given in less detail. To derive the backward equation, we consider the Chapman-Kolmogorov equation corresponding to a transition from time $\tau - \Delta\tau$, $\Delta\tau > 0$, to time $\tau$ and then from time $\tau$ to time $t$. The equation required is

$$P(\tau - \Delta\tau, x; t, A) = \int P(\tau - \Delta\tau, x; \tau, dy)P(\tau,y;t,A). \qquad (5)$$

A small time transition is considered at the beginning $\tau - \Delta\tau$ rather than at the end $t$ (as in the derivation of the forward equation). This, in part, explains the names "backward" and "forward" equations. If we split the range of integration on the right of (5) into $\Omega_X - \{x\}$ and $\{x\}$, we obtain

$$P(\tau - \Delta\tau, x; t, A) - P(\tau - \Delta\tau, x; \tau, \{x\})P(\tau,x;t,A)$$
$$= [P(\tau - \Delta\tau, x; t, A) - P(\tau,x;t,A)] + p(\tau,x)\Delta\tau P(\tau,x;t,A) + o(\Delta\tau)$$
$$= \int_{\Omega_X - \{x\}} P(\tau,y;t,A)P(\tau - \Delta\tau, x; \tau, dy) \qquad (6)$$

using the basic property (1) of jump processes. For every set $A$ not containing $x$, $P(\tau - \Delta\tau, x; \tau, A)/\Delta\tau$ approaches $p(\tau,x)\Pi(\tau,x,A)$ as $\Delta\tau \to 0$. On dividing equation (6) by $\Delta\tau$ and letting $\Delta\tau \to 0$, the following identity is obtained

$$\frac{\partial P(\tau,x;t,A)}{\partial \tau} = p(\tau,x)\left\{ P(\tau,x;t,A) - \int P(\tau,y;t,A)\Pi(\tau,x,dy) \right\} \qquad (7)$$

where the derivative on the left is to be understood as a left-hand derivative. By considering a transition from $\tau$ to $\tau + \Delta\tau$ and then from $\tau + \Delta\tau$ to $t$, the corresponding equation is obtained with a right-hand derivative using the continuity of $p(t,x)$ and $\Pi(t,x,A)$. Thus, the equation holds with $\partial P/\partial\tau$ understood as an ordinary derivative.

The forward equation is obtained by considering a transition from $\tau$ to $t$ and then from $t$ to $t + \Delta t$ in the Chapman-Kolmogorov equation

$$P(\tau, x; t + \Delta t, A) = \int P(t, y; t + \Delta t, A)P(\tau,x;t,dy). \qquad (8)$$

A more convenient form in which to write this is

$$\frac{1}{\Delta t}\{P(\tau, x; t + \Delta t, A) - P(\tau,x;t,A)\}$$
$$= \int P(\tau,x;t,dy)\{P(t, y; t + \Delta t, A) - \delta(y,A)\}/\Delta t \qquad (9)$$

and the desired result (4) follows from this on letting $\Delta t \to 0$ by (1) and bounded convergence. Actually, without the boundedness condition on $p(t,x)$, one would not get an equality in the forward relation, only an inequality.

It has been established that the transition probability function satisfying (1), if it exists, should satisfy the backward and forward equations. Two basic questions now arise, that of existence of a solution and if it exists whether it is unique. Our object is now to construct a solution of the problem and show that it is unique. Let

$$P_0(\tau,x;t,A) = \delta(x,A) \exp\left\{-\int_\tau^t p(s,x)\,ds\right\}. \tag{10}$$

This is the probability of no jump in the time interval $(\tau,t)$. We now construct the probability of precisely $n$ jumps in the time interval $(\tau,t)$. Introduce the function

$$\Pi^*(\tau,x;t,A) = \int_A \exp\left\{-\int_\tau^t p(s,y)\,ds\right\} \Pi(\tau,x,dy). \tag{11}$$

We shall try to construct the desired probability recursively. Assume that the probability of $n-1$ jumps precisely in time $(\tau,t)$ in going from $x$ at time $\tau$ into set $A$ at time $t$, $P_{n-1}(\tau,x;t,A)$, is already given. Intuitively, it would follow that one ought to be able to construct $P_n(\tau,x;t,A)$ out of $P_{n-1}(\tau,x;t,A)$ in the following way. If precisely $n$ jumps have taken place in $(\tau,t)$, the $n$-th or last jump must have taken place at precisely some intermediate time $\sigma$, $\tau < \sigma < t$. There will then be $n-1$ jumps in the open interval $(\tau,\sigma)$ and none in the open interval $(\sigma,t)$. Thus

$$P_n(\tau,x;t,A) = \int_\tau^t d\sigma \int_{\Omega x} p(\sigma,y)\Pi^*(\sigma,y;t,A)P_{n-1}(\tau,x;\sigma,dy). \tag{12}$$

A similar argument in terms of the first jump instead of the $n$-th jump leads to the equation

$$P_n(\tau,x;t,A) = \int_\tau^t p(\sigma,x) \exp\left\{-\int_\tau^\sigma p(s,x)\,ds\right\}. \tag{13}$$
$$\int P_{n-1}(\sigma,y;t,A)\Pi(\sigma,x,dy)\,d\sigma.$$

The functions $P_n(\tau,x;t,A)$ as generated from $P_0(\tau,x;t,A)$ by either system (12) or system (13) can be seen to be the same on writing them out as iterated integrals. Of course, we have been guided in our definition of the functions $P_n$ by intuitive considerations. But we have to verify that they are in fact probabilities. Let

$$S_n(\tau,x;t,A) = \sum_{j=0}^n P_j(\tau,x;t,A). \tag{14}$$

Then by (13)

$$S_n(\tau,x;t,A) = \exp\left\{-\int_\tau^t p(s,x)\,ds\right\}\left\{\delta(x,A)\right. \tag{15}$$
$$\left. + \int_\tau^t p(\sigma,x)\exp\left\{\int_\sigma^t p(s,x)\,ds\right\}d\sigma\int S_{n-1}(\sigma,y;t,A)\Pi(\sigma,x,dy)\right\}.$$

Now $0 \leq S_0(\tau,x;t,A) = P_0(\tau,x;t,A) \leq 1$. Further $S_n \geq S_{n-1} \geq 0$. If $S_{n-1} \leq 1$, by equation (15)

$$S_n(\tau,x;t,A) \leq \exp\left\{-\int_\tau^t p(s,x)\,ds\right\}$$
$$\left\{1 + \int_\tau^t p(\sigma,x)\exp\left\{\int_\sigma^t p(s,x)\,ds\right\}d\sigma\right\} = 1. \quad (16)$$

By induction the sequence $S_n(\tau,x;t,A)$ is a nondecreasing sequence bounded below by zero and above by one. Thus

$$P(\tau,x;t,A) = \sum_{j=0}^{\infty} P_j(\tau,x;t,A) = \lim_{n\to\infty} S_n(\tau,x;t,A) \leq 1 \quad (17)$$

is well defined. The term $P_n(\tau,x;t,A)$ can therefore be interpreted as the conditional probability of going from $x$ at time $\tau$ into $A$ at time $t$ with precisely $n$ jumps or changes of state. It is still an open point as to just when $P(\tau,x;t,\Omega_x) \equiv 1$. $P(\tau,x;t,A)$ is the conditional probability of going from $x$ at time $\tau$ into $A$ at time $t$ in a finite number of jumps.

Let

$$L_n(\tau,x,t) = \int_\tau^t d\sigma \int p(\sigma,y)P_n(\tau,x;\sigma,dy) \quad (18)$$

be the conditional probability of having had at least $n+1$ jumps in time $(\tau,t)$ given that the system was in state $x$ at time $\tau$. Notice that

$$\frac{\partial \Pi^*(\tau,x;t,A)}{\partial t} = -\int_A p(t,y)\Pi^*(\tau,x;t,dy). \quad (19)$$

It follows from (12) and (19) that

$$\int_\tau^t d\sigma_1 \int_A p(\sigma_1,y)P_{n+1}(\tau,x;\sigma_1,dy)$$
$$= -\int_\tau^t d\sigma_1 \int_\tau^{\sigma_1} d\sigma \int p(\sigma,y) \frac{\partial \Pi^*(\sigma,y;\sigma_1,A)}{\partial \sigma_1} P_n(\tau,x;\sigma,dy). \quad (20)$$

An interchange of order of integration and the observation that $\Pi^*(\sigma,y;\sigma,A) = \Pi(\sigma,y,A)$ leads us to the equation

$$\int_\tau^t d\sigma_1 \int_A p(\sigma_1,y)P_{n+1}(\tau,x;\sigma_1,dy)$$
$$= \int_\tau^t d\sigma \int p(\sigma,y)\Pi(\sigma,y,A)P_n(\tau,x;\sigma,dy)$$
$$- \int_\tau^t d\sigma \int p(\sigma,y)\Pi^*(\sigma,y;t,A)P_n(\tau,x;\sigma,dy). \quad (21)$$

This can be recast in the more convenient form

$$P_{n+1}(\tau,x;t,A) + \int_\tau^t d\sigma \int_A p(\sigma,y)P_{n+1}(\tau,x;\sigma,dy)$$
$$= \int_\tau^t d\sigma \int p(\sigma,y)\Pi(\sigma,y,A)P_n(\tau,x;\sigma,dy) \qquad (22)$$

by using (12). The interesting relation

$$P_{n+1}(\tau,x;t,\Omega_x) + L_{n+1}(\tau,x,t) = L_n(\tau,x,t) \qquad (23)$$

follows from (22) on setting $A = \Omega_x$. Now

$$P_0(\tau,x;t,\Omega_x) + L_0(\tau,x,t) = 1. \qquad (24)$$

This coupled with (23) indicates that

$$P(\tau,x;t,\Omega_x) = \lim_{N\to\infty} \sum_{n=0}^N P_n(\tau,x;t,\Omega_x)$$
$$= \lim_{N\to\infty} (1 - L_N(\tau,x,t)) \qquad (25)$$

equals one if and only if the limit $L(\tau,x,t)$ of the nonincreasing sequence $L_n(\tau,x,t)$ is zero. But

$$L(\tau,x,t) \le \int_\tau^t d\sigma \int p(\sigma,y)P_n(\tau,x;\sigma,dy)$$
$$\le K \int_\tau^t P_n(\tau,x;\sigma,\Omega_x)\,d\sigma \qquad (26)$$

since $p(\sigma,y)$ is bounded by a constant $K$. The right-hand side of (26) is the general term of a convergent series so that $L(\tau,x,t)$ must be zero. It follows that $P(\tau,x;t,A)$ is a probability measure in $A$.

From the definition it follows that for $\tau < \lambda < t$

$$P_n(\tau,x;t,A) = \sum_{k=0}^n \int P_k(\tau,x;\lambda,dy)P_{n-k}(\lambda,y;t,A). \qquad (27)$$

Of course this only states that if there are precisely $n$ jumps in going from $x$ into $A$ in $(\tau,t)$, for some integer $k$ $(k = 0, 1, \ldots, n)$ there will be $k$ jumps in $(\tau,\lambda)$ and $n - k$ jumps in $(\lambda,t)$. On summing over $n$ the Chapman-Kolmogorov equation (a.1) is obtained. The construction of the functions $P_n(\tau,x;t,A)$ (see (13)) and the boundedness of $p(t,x)$ imply that $P(\tau,x;t,A)$ satisfies condition (1). The arguments given earlier in this section imply that the function $P(\tau,x;t,A)$ constructed must satisfy both the forward and backward equations.

Let us now consider the uniqueness of $P(\tau,x;t,A)$. Suppose there were another transition probability function $P^*(\tau,x;t,A)$ satisfying the Chapman-Kolmogorov equation and condition (1). $P^*(\tau,x;t,A)$ would have to satisfy the backward equation (3). If the backward equation is treated as an ordinary differential equation, the following equation is obtained

$$P^*(\tau,x;t,A) = \exp\left\{-\int_\tau^t p(s,x)\,ds\right\}\left\{\delta(x,A)\right. \tag{28}$$
$$\left. + \int_\tau^t p(\sigma,x)\exp\left\{-\int_\sigma^t p(s,x)\,ds\right\}d\sigma\int P^*(\sigma,y;t,A)\Pi(\sigma,x,dy)\right\}.$$

Clearly $P^*(\tau,x;t,A) \geq P_0(\tau,x;t,A) = S_0(\tau,x;t,A)$. But this implies that $P^*(\tau,x;t,A) \geq S_1(\tau,x;t,A)$. By iterating this argument we see that $P^*(\tau,x;t,A) \geq S_n(\tau,x;t,A)$ for every $n$. Therefore $P^*(\tau,x;t,A) \geq P(\tau,x;t,A)$. But the inequality must be equality since $P(\tau,x;t,A)$ is already a probability measure in $A$.

The boundedness condition on $p(t,x)$ (see condition (i)) implies that transitions over a finite time period take place through a finite number of jumps. To a great extent, what we have shown above amounts to this apparently simple statement. If the boundedness condition is relaxed, transitions can take place in a much more complicated manner. There may no longer be a finite number of jumps with probability one in a finite time interval. In fact, the set of time points where jumps occur may have many limit points. Further, uniqueness of a transition probability function satisfying the infinitesimal condition (1) no longer follows. The more complicated behavior of sample functions and the possible nonuniqueness of solutions are two related aspects of the difficulties that arise when one allows for unbounded or even infinite $p(t,x)$. Much of the recent work on Markov processes has been concerned with problems of this type (see Chung [8]).

The Markov processes with stationary transition mechanism have been investigated most extensively. If such a Markov process has a countable state space, we refer to it as a Markov chain just as in the case of a discrete time parameter. There are a few simple remarks that can be made about finite state Markov chains with stationary transition mechanism. The transition probability function of a Markov chain with stationary transition mechanism can be conveniently written in matrix form

$$\mathbf{P}(t) = (p_{i,j}(t); i,j = 1, 2, \ldots), t \geq 0, \tag{29}$$
$$p_{i,j}(t) = P[X(\tau + t) = j | X(\tau) = i].$$

The Chapman-Kolmogorov equation is then

$$\mathbf{P}(t)\mathbf{P}(s) = \mathbf{P}(t + s), t, s \geq 0. \tag{30}$$

If $\mathbf{P}(t)$ corresponds to a "decent" Markov chain, it will be continuous with

$$\lim_{t\to 0+} \mathbf{P}(t) = \mathbf{I}. \tag{31}$$

For sufficiently small $t$, $\mathbf{P}(t)$ will be close to $\mathbf{I}$. More formally, given any $\varepsilon > 0$, there is a $\delta(\varepsilon) > 0$ such that for $0 \le t < \delta(\varepsilon)$, $|p_{i,j}(t) - \delta_{i,j}| < \varepsilon$, $i, j = 1, \ldots, n$, where $n$ is the number of states of the chain. For sufficiently small $\varepsilon > 0$, log $\mathbf{A}$ is well defined for matrices $\mathbf{A}$ within the $\varepsilon$ neighborhood of $I$, $|a_{i,j} - \delta_{i,j}| < \varepsilon$, $i, j = 1, \ldots, n$, by

$$\log \mathbf{A} = \log (\mathbf{I} - (\mathbf{I} - \mathbf{A})) = -\sum_{k=1}^{\infty} (\mathbf{I} - \mathbf{A})^k / k. \tag{32}$$

As expected exp $\{\log \mathbf{A}\} = \mathbf{A}$. Further, for commuting matrices $\mathbf{A}$, $\mathbf{B}$ such that $\mathbf{A}$, $\mathbf{B}$, $\mathbf{AB}$ are in the $\varepsilon$ neighborhood of $\mathbf{I}$ spoken of above

$$\log \mathbf{AB} = \log \mathbf{A} + \log \mathbf{B}. \tag{33}$$

The Chapman-Kolmogorov equation indicates that the matrices $\mathbf{P}(t)$ commute with each other. Thus, for sufficiently small $t$,

$$\log \mathbf{P}(t + \tau) = \log \mathbf{P}(t)\mathbf{P}(\tau) = \log \mathbf{P}(t) + \log \mathbf{P}(\tau). \tag{34}$$

But equation (34) will hold for continuous $\mathbf{P}(t)$ and $t$ sufficiently small if and only if

$$\log \mathbf{P}(t) = \mathbf{B}t \tag{35}$$

where $\mathbf{B}$ is an $n \times n$ matrix. Thus $\mathbf{P}(t) = \exp(\mathbf{B}t)$ for $t$ sufficiently small. The Chapman-Kolmogorov equation implies that $\mathbf{P}(t) = [\mathbf{P}(t/n)]^n$ for all positive integral $n$ so that

$$\mathbf{P}(t) = \exp(\mathbf{B}t) \tag{36}$$

for all $t \ge 0$. Notice that $\mathbf{B} = \mathbf{P}'(0)$, the derivative of $\mathbf{P}(t)$ at $t = 0$. Since $\mathbf{P}(t)$ is a transition probability matrix with $\mathbf{P}(0) = \mathbf{I}$, it follows that

$$b_{i,i} = p'_{i,i}(0) = \lim_{t\to 0+} \frac{p_{i,i}(t) - 1}{t} \le 0$$

$$b_{i,j} = p'_{i,j}(0) = \lim_{t\to 0+} \frac{p_{i,j}(t)}{t} \ge 0, j \ne i, \tag{37}$$

$$\sum_j b_{i,j} = \sum_j p'_{i,j}(0) = \frac{d}{dt} \sum_j p_{i,j}(t)\Big|_{t=0} = \frac{d}{dt}(1) = 0.$$

It is interesting that the assumption that $\mathbf{P}(t)$ is a transition probability matrix of finite order $n$, continuous with $\lim_{t \to 0+} \mathbf{P}(t) = \mathbf{I}$, is enough to imply that conditions (i), (ii) cited at the beginning of section b are satisfied in the form (37). The argument used in this section to construct transition probability functions in the case of a general state space shows that any matrix $\mathbf{B}$ of order $n$ with

$$
\begin{aligned}
&b_{i,i} \leq 0 \\
&b_{i,j} \geq 0, \quad j \neq i, \\
&\sum_j b_{i,j} = 0
\end{aligned}
\tag{38}
$$

can be used to construct a transition probability matrix $\mathbf{P}(t) = \exp(\mathbf{B}t)$. Thus, the general form of a continuous transition probability matrix with $\mathbf{P}(t) \to \mathbf{I}$ as $t \to 0$ has been given in the case of a finite state space. The backward and forward equations assume a particularly simple form. The backward equation is simply

$$
\frac{d\mathbf{P}(t)}{dt} = \mathbf{B}\mathbf{P}(t)
\tag{39}
$$

or

$$
\frac{dp_{i,j}(t)}{dt} = \sum_k b_{i,k} p_{k,j}(t), \quad i, j = 1, \ldots, n,
\tag{40}
$$

while the forward equation is given by

$$
\frac{d\mathbf{P}(t)}{dt} = \mathbf{P}(t)\mathbf{B}
\tag{41}
$$

or

$$
\frac{dp_{i,j}(t)}{dt} = \sum_k p_{i,k}(t) b_{k,j}, \quad i, j = 1, \ldots, n.
\tag{42}
$$

As already remarked, the situation in the case of a denumerable state space is much more complicated. In fact, many questions are as yet unresolved. An extensive discussion of the current state of knowledge on Markov chains with stationary transition mechanism and an infinite number of states is given in Chung's monograph [8].

There are a number of types of Markov chains that have been investigated in some detail. We briefly discuss one such type, the "birth and death" processes. These processes are often taken as models of population development through time, as might be inferred from their name. Let us think of $j$, $j \geq 0$, as the population size. The state $j = 0$ corresponds to the death of the population, that is, no individuals at all in the population. In a small time, if there is any change in population size at all, it is either through one birth or one death. Thus

$$p'_{0,j}(0) = b_{0,j} = 0$$
$$p'_{i,j}(0) = b_{i,j} = 0 \text{ if } j \neq i, i - 1, i + 1 \tag{43}$$

with

$$\sum_j b_{i,j} = 0. \tag{44}$$

An extensive discussion of the birth and death processes has been carried out in a series of papers of Karlin and MacGregor [39]. Their work is based on the observation that structural questions concerning these processes are closely related to classical moment problems.

## c. Diffusion Processes

We have already noted that the jump processes are roughly characterized by the fact that in a short time the system will with large probability not change state, but that if it does move it will by an appreciable amount. Of course, this was not completely apparent in the previous section since a topology (the concept of neighborhood or a notion of closeness) was not introduced in the discussion because strictly speaking it was not required. The diffusion processes are at the other extreme. A topology is necessary here explicitly or implicitly. In a rough and inexact way the diffusion processes are characterized by the property that in a short time one is sure that the system will move, but only by a small amount. The Wiener process that was introduced in section d of Chapter IV is the simplest and the basic example of a diffusion process.

We shall give a limited discussion of diffusion processes on the real line. Let $F(\tau,\xi;t,x)$ be the conditional distribution function

$$F(\tau,\xi;t,x) = P[X(t) \leq x | X(\tau) = \xi], \tau < t. \tag{1}$$

Since the process $X(t)$ is assumed to be Markovian, the conditional distribution function $F(\tau,\xi;t,x)$ satisfies the Chapman-Kolmogorov equation

$$F(\tau,\xi;t,x) = \int_{-\infty}^{\infty} F(\sigma,y;t,x) d_y F(\tau,\xi;\sigma,y), \tau < \sigma < t. \tag{2}$$

The function $F(\tau,\xi;t,x)$ determines the full probability structure of the process $X(t)$ since the process is Markovian.

We shall call the process $X(t)$ a diffusion process if it satisfies the following conditions

$$\lim_{\Delta t \to 0} \frac{1}{\Delta t} \int_{|y-x| \geq \delta} d_y F(t - \Delta t, x; t, y) = 0 \tag{3}$$

$$\lim_{\Delta t \to 0} \frac{1}{\Delta t} \cdot \int_{|y-x| < \delta} (y - x)^2 d_y F(t - \Delta t, x; t, y) = a(t,x) \geq 0 \tag{4}$$

$$\lim_{\Delta t \to 0} \frac{1}{\Delta t} \int_{|y-x| < \delta} (y - x) d_y F(t - \Delta t, x; t, y) = b(t,x) \tag{5}$$

for any fixed positive δ. The first limit condition (3) states that the probability of drifting away from an initial position $x$ by an amount greater than δ in time $\Delta t$ is of smaller order than $\Delta t$ for $\Delta t$ small. Thus, in a small time interval the particle in diffusion (if we regard $X(t)$ as the position of the diffusing particle at time $t$) is almost sure to remain in the immediate neighborhood of its initial position. The relations (4) and (5) are conditions on the truncated variance and mean displacement of the particle in a small time $\Delta t$. Notice that the Wiener process satisfies these conditions with $a(t,x) \equiv 1$ and $b(t,x) = 0$. This suggests that a diffusion process $X(t)$ satisfying conditions (3), (4), and (5) could be regarded as acting locally in time like a Wiener process $Y(t)$ in that the change in position $\Delta X(t) = X(t + \Delta t) - X(t)$ is given to the first order by

$$b(t,X(t))\, \Delta t + [a(t,X(t))]^{\frac{1}{2}}\, \Delta Y(t) \tag{6}$$

where $\Delta Y(t) = Y(t + \Delta t) - Y(t)$. This idea appears to be due to S. Bernstein [4] and was used very effectively by K. Ito [36]. This approach involves direct analysis of the stochastic process. Our interest is basically analytic and it will therefore concern itself with the analytic properties of the conditional distribution function $F(\tau,\xi;t,x)$.

Assume that the partial derivatives

$$\frac{\partial F(\tau,\xi;t,x)}{\partial \xi}, \frac{\partial^2 F(\tau,\xi;t,x)}{\partial \xi^2} \tag{7}$$

exist. Using a Taylor expansion with error term, we find that

$$\frac{F(\tau - \Delta\tau, \xi; t, x) - F(\tau,\xi;t,x)}{\Delta\tau}$$

$$= \frac{1}{\Delta\tau} \int\limits_{|y-\xi|>\delta} \{F(\tau,y;t,x) - F(\tau,\xi;t,x)\} d_y F(\tau - \Delta\tau, \xi; \tau, y)$$

$$+ \frac{\partial F(\tau,\xi;t,x)}{\partial\xi} \frac{1}{\Delta\tau} \int\limits_{|y-\xi|<\delta} (y - \xi) d_y F(\tau - \Delta\tau, \xi; \tau, y) \qquad (8)$$

$$+ \frac{\partial^2 F(\tau,\xi;t,x)}{\partial\xi^2} \frac{1}{\Delta\tau} \int\limits_{|y-\xi|<\delta} \left\{\frac{(y-\xi)^2}{2} + o(y-\xi)^2\right\} d_y F(\tau - \Delta\tau, \xi; \tau, y).$$

On letting $\Delta\tau \to 0$ and making use of conditions (3), (4) and (5), one finds that a left-hand derivative

$$\frac{\partial F(\tau,\xi;t,y)}{\partial\tau} = - \lim_{\Delta\tau\to 0} \frac{F(\tau - \Delta\tau, \xi; t, x) - F(\tau,\xi;t,x)}{\Delta\tau} \qquad (9)$$

exists and satisfies the partial differential equation

$$\frac{\partial F}{\partial\tau} + \frac{1}{2} a(\tau,\xi) \frac{\partial^2 F}{\partial\xi^2} + b(\tau,\xi) \frac{\partial F}{\partial\xi} = 0. \qquad (10)$$

This differential equation is a parabolic differential equation (see [74]) in the backward variables $\tau$, $\xi$. This suggests that one ought to look for solutions of equation (10) that have the properties of a conditional distribution function and satisfy the Chapman-Kolmogorov equation. W. Feller has in fact done this under a variety of boundedness and regularity conditions on the coefficients $a(\tau,\xi)$, $b(\tau,\xi)$ of the equation (10). Notice that if $a(\tau,\xi)$ is bounded away from zero and infinity and infinity and differentiable in $\xi$, the equation (10) can be reduced to an equation of the form

$$\frac{\partial G}{\partial\tau} + \frac{1}{2} \frac{\partial^2 G}{\partial\eta^2} + c(\tau,\eta) \frac{\partial G}{\partial\eta} = 0 \qquad (11)$$

by the transformation

$$\eta = \int_0^\xi \frac{dy}{\sqrt{a(\tau,y)}} \qquad (12)$$

(see Problem 8 of this chapter). This implies that one might just as well consider the simpler equation

$$\frac{\partial F}{\partial \tau} + \frac{1}{2}\frac{\partial^2 F}{\partial \xi^2} + b(\tau,\xi)\frac{\partial F}{\partial \xi} = 0. \tag{13}$$

If $F$ has a sufficiently smooth density function $f$

$$F(\tau,\xi;t,x) = \int_{-\infty}^{x} f(\tau,\xi;t,y)\, dy, \tag{14}$$

the density function $f(\tau,\xi;t,x)$ will satisfy the same differential equation

$$\frac{\partial f}{\partial \tau} + \frac{1}{2}\frac{\partial^2 f}{\partial \xi^2} + b(\tau,\xi)\frac{\partial f}{\partial \xi} = 0. \tag{15}$$

We shall briefly indicate how one might construct a solution $f$ of (15) assuming that $b(\tau,\xi)$ is sufficiently smooth. The argument is a heuristic version of that used by Feller in his paper [15].

First notice that if $b(\tau,\xi)$ were identically zero, the transition probability density of the Wiener process

$$u(\tau,\xi;t,x) = [2\pi(t - \tau)]^{-\frac{1}{2}} \exp \{-\frac{1}{2}(x - \xi)^2/(t - \tau)\} \tag{16}$$

would satisfy the differential equation (15) and would have the desired properties. Let

$$G(\tau,\xi) = \int_{\tau}^{T} dt \int_{-\infty}^{\infty} g(t,x)u(\tau,\xi;t,x)\, dx. \tag{17}$$

If $g(t,x)$ satisfies a Hölder condition of the form

$$|g(t,x) - g(t',x')| < K\{|t - t'|^{\alpha} + |x - x'|^{\alpha}\}, \alpha > 0 \tag{18}$$

in a neighborhood of every point $(t,x)$, one can show that the partial derivatives $\dfrac{\partial G}{\partial \tau}, \dfrac{\partial G}{\partial \xi}, \dfrac{\partial^2 G}{\partial \xi^2}$ of $G$ exist and that $G$ satisfies the partial differential equation

$$\frac{\partial G}{\partial \tau} + \frac{1}{2}\frac{\partial^2 G}{\partial \xi^2} = -g(\tau,\xi) \tag{19}$$

(see Problem 9 of this chapter). Consider now using the following iterative procedure. Let

$$f_0(\tau,\xi;t,x) = u(\tau,\xi;t,x)$$

$$f_{n+1}(\tau,\xi;t,x) = \int_{\tau}^{t} dp \int_{-\infty}^{\infty} b(p,q)\frac{\partial f_n(p,q;t,x)}{\partial q} f_0(\tau,\xi;p,q)\, dq \tag{20}$$

$$n = 0, 1, \ldots.$$

With sufficiently strong regularity conditions on the function $b(\tau,\xi)$ one should be able to show that all the functions $f_n$ are well defined and smooth and that

$$\frac{\partial f_{n+1}}{\partial \tau} + \frac{1}{2} \frac{\partial^2 f_{n+1}}{\partial \xi^2} = -b(\tau,\xi) \frac{\partial f_n}{\partial \xi}. \tag{21}$$

If the series $\Sigma f_n$ and the appropriate series in derivatives of $f_n$ converge rapidly enough

$$f = \sum_{n=0}^{\infty} f_n \tag{22}$$

will be a solution of the differential equation (15). Of course, this solution is one obtained by a successive approximation. Feller [15] has also shown that the function $f(\tau,\xi;t,x)$ constructed above also satisfies the partial differential equation

$$\frac{\partial f}{\partial t} = \frac{1}{2} \frac{\partial^2 f}{\partial x^2} - \frac{\partial}{\partial x} (b(t,x)f)$$

(commonly called the forward equation) in the forward variables $t$, $x$ under the boundedness and regularity conditions on $a(t,x)$, $b(t,x)$ referred to. If $a(t,x) \neq 1$ the partial differential equation

$$\frac{\partial f}{\partial t} = \frac{1}{2} \frac{\partial^2}{\partial x^2} (a(t,x)f) - \frac{\partial}{\partial x} (b(t,x)f)$$

would be satisfied in the forward variables. This differential equation is sometimes referred to as a Fokker-Planck equation [18].

## d. A Refined Model of Brownian Motion

We have already remarked that the Gaussian process $X(t)$, $0 \leq t < \infty$, with $X(0) \equiv 0$, mean $EX(t) \equiv 0$ and covariance function $EX(t)X(\tau) = \min(t,\tau)$ has been used to represent the position of a particle in Brownian motion at time $t$. Here, the diffusion constant has been set equal to unity for convenience. The process is Markovian and its transition probability density

$$p(t - \tau, y - x) = \frac{\partial}{\partial y} P[X(t) \leq y | X(\tau) = x]$$
$$= [2\pi(t - \tau)]^{-\frac{1}{2}} \exp \{-(y - x)^2/2(t - \tau)\} \tag{1}$$

satisfies the forward differential equation

$$\frac{\partial p}{\partial t} = \frac{1}{2}\frac{\partial^2 p}{\partial y^2} \tag{2}$$

in $y$, $t$ and the backward differential equation

$$-\frac{\partial p}{\partial \tau} = \frac{1}{2}\frac{\partial^2 p}{\partial x^2} \tag{3}$$

in $x$, $\tau$. However, this model is unpleasant in that it leads to nondifferentiable sample functions, that is, the particle in Brownian motion has no well-defined velocity. This is suggested by the fact that the difference quotient

$$\frac{X(t+h) - X(t)}{h} \tag{4}$$

does not converge in mean square as $h \to 0$. In fact, it is readily verified that the second moment of the difference quotient (4) diverges as $h \to 0$.

Ornstein and Uhlenbeck [58] constructed a model of Brownian motion in which the particle in diffusion has a well-defined velocity. We first introduce the random process $U(t)$ describing the velocity of the particle. $U(t)$ is given in terms of the process $X(t)$ as follows

$$U(t) = \sqrt{\alpha}\,\exp\,(-\beta t)X(e^{2\beta t}), \quad -\infty < t < \infty. \tag{5}$$

It is clear that $U(t)$ is a Gaussian process since it is derived from the Gaussian process $X(t)$ by (5). The first and second moment properties of $U(t)$ are therefore enough to characterize the full probability structure of the process. The first moment is identically zero since

$$EU(t) = \sqrt{\alpha}\,\exp\,(-\beta t)EX(e^{2\beta t}) \equiv 0 \tag{6}$$

and the covariance function is given by

$$\begin{aligned} EU(t)U(\tau) &= \alpha \exp\,(-\beta(t+\tau))EX(e^{2\beta t})X(e^{2\beta \tau}) \\ &= \alpha \exp\,(-\beta(t+\tau))\,\exp\,(2\beta \min\,(t,\tau)) \\ &= \alpha \exp\,(-\beta|t-\tau|). \end{aligned} \tag{7}$$

Thus $U(t)$ is a strictly stationary process. In fact, $U(t)$ is just $X(t)$ appropriately modified so as to make it stationary. This has been accomplished by a change of scale in time so as to take zero into $-\infty$ and a renormalization by $\exp\,(-\beta t)$. Further $U(t)$ is *Markovian* since it is obtained from $X(t)$ by an instantaneous one-one transformation and

$X(t)$ is Markovian. The process $U(t)$ is usually called the Ornstein-Uhlenbeck process. Notice that the Ornstein-Uhlenbeck process satisfies the formal equation

$$dU(t) + \beta U(t)\, dt = \sqrt{\alpha}\, \exp{(-\beta t)}\, dX(e^{2\beta t}) \approx \sqrt{2\alpha\beta}\, dX(t). \quad (8)$$

Here $\sqrt{2\alpha\beta}\, dX(t)$ can be regarded as a random impulse where $2\alpha\beta$ is a measure of the mean square displacement of the impulse and $\beta$ is a friction coefficient. We shall set $\alpha = 1$ for convenience. The transition probability density of $U(t)$ can be written down readily using the properties of $X(t)$. Since the increments of $X(t)$ over nonoverlapping intervals are independent, it follows from (5) that

$$U(t) - \exp{(-\beta(t - \tau))}U(\tau), \quad t > \tau, \quad (9)$$

is independent of $U(\tau)$. Therefore the conditional expected value

$$
\begin{aligned}
E[U(t)|U(\tau)] &= E[U(t) - \exp{(-\beta(t - \tau))}U(\tau)|U(\tau)] \\
&\quad + \exp{(-\beta(t - \tau))}U(\tau) \\
&= \exp{(-\beta(t - \tau))}U(\tau) \quad (10)
\end{aligned}
$$

and by formula (7)

$$
\begin{aligned}
E[\{U(t) &- \exp{(-\beta(t - \tau))}U(\tau)\}^2|U(\tau)] \\
&= E[U(t) - \exp{(-\beta(t - \tau))}U(\tau)]^2 \\
&= 1 - \exp{(-2\beta(t - \tau))}, \quad t > \tau. \quad (11)
\end{aligned}
$$

The conditional distribution function of $U(t)$ given $U(\tau)$ is Gaussian since the process $U(t)$ is Gaussian. The conditional mean and variance of $U(t)$ given $U(\tau)$ are (10) and (11) respectively. It then follows that the transition probability density

$$
\begin{aligned}
p(\tau,x;t,y) &= \frac{\partial}{\partial y} P[U(t) \leq y|U(\tau) = x] \\
&= (2\pi)^{-\frac{1}{2}}[1 - \exp{(-2\beta(t - \tau))}]^{-\frac{1}{2}} \quad (12) \\
&\quad \exp{\{-[y - \exp{(-\beta(t - \tau))}x]^2/2[1 - \exp{(-2\beta(t - \tau))}]\}}.
\end{aligned}
$$

Further, notice that

$$
\begin{aligned}
E[U(t + h) - U(t)|U(t)] &= \exp{(-\beta h)}U(t) - U(t) \\
&= -\beta h U(t) + o(h) \quad (13)
\end{aligned}
$$

and

$$
\begin{aligned}
E[\{U(t + h) &- U(t)\}^2|U(t)] \\
&= E[\{U(t + h) - \exp{(-\beta h)}U(t) + U(t)(\exp{(-\beta h)} - 1)\}^2|U(t)] \\
&= 1 - \exp{(-2\beta h)} + U(t)^2(\exp{(-\beta h)} - 1)^2 \\
&= 2\beta h + o(h). \quad (14)
\end{aligned}
$$

This indicates that the transition probability density satisfies the forward equation

$$\frac{\partial p}{\partial t} = \beta \frac{\partial^2 p}{\partial y^2} + \beta \frac{\partial}{\partial y} (yp) \tag{15}$$

in $y$, $t$ and the backward equation

$$-\frac{\partial p}{\partial \tau} = \beta \frac{\partial^2 p}{\partial x^2} - \beta x \frac{\partial p}{\partial x} \tag{16}$$

in $x$, $\tau$. Of course, this could be verified directly.

If the velocity of the particle is to be described by $U(t)$, the position of the particle must be given by

$$B(t) = \int_0^t U(a) \, da \tag{17}$$

assuming that the particle starts at zero at time $t = 0$. The process $B(t)$ is Gaussian since it is derived from a Gaussian process by a linear operation. The mean

$$EB(t) = \int_0^t EU(a) \, da \equiv 0 \tag{18}$$

and the covariance

$$
\begin{aligned}
EB(t)B(\tau) &= \int_0^t \int_0^\tau EU(a)U(b) \, da \, db \\
&= \int_0^t \int_0^\tau \exp\left(-\beta|a - b|\right) da \, db \\
&= \int_{\min(t,\tau)}^{\max(t,\tau)} \int_0^{\min(t,\tau)} \exp\left(-\beta(a - b)\right) db \, da \\
&\quad + 2 \int_0^{\min(t,\tau)} \int_0^a \exp\left(-\beta(a - b)\right) db \, da \\
&= \frac{2}{\beta} \min(t,\tau) + \frac{1}{\beta^2} [\exp\left(-\beta \min(t,\tau)\right) \\
&\quad + \exp\left(-\beta \max(t,\tau)\right) - \exp\left(-\beta|t - \tau|\right) - 1].
\end{aligned}
\tag{19}
$$

Clearly if we set $\beta = 2$, for large $t$, $\tau$ and $|t - \tau|$ the process $B(t)$ looks very much like the process $X(t)$. However, the process $B(t)$ is *not Markovian. If $B(t)$ is considered jointly with $U(t)$, a two-dimensional Markov process is obtained.* Let us see what the equation satisfied by the transition probability density looks like. By (13) and (14)

$$E[U(t + h) - U(t)|U(t), B(t)] = -\beta h U(t) + o(h) \tag{20}$$

and

$$E[\{U(t + h) - U(t)\}^2|U(t), B(t)] = 2\beta h + o(h). \tag{21}$$

Further

$$E[B(t + h) - B(t)|U(t), B(t)] = E\left[\int_t^{t+h} U(a)\, da|U(t)\right] \qquad (22)$$

$$= E\left[\int_t^{t+h} \{U(a) - \exp\left(-\beta(a - t)\right)U(t)\}\, da|U(t)\right]$$

$$+ \int_t^{t+h} \exp\left(-\beta(a - t)\right) da\ U(t)$$

$$= hU(t) + o(h)$$

while

$$E[\{B(t + h) - B(t)\}^2|U(t), B(t)]$$

$$= E\left[\left\{\int_t^{t+h} U(a)\, da\right\}^2|U(t)\right] = O(h^2). \qquad (23)$$

Notice that (21) and (23) imply that

$$E[\{B(t + h) - B(t)\}\{U(t + h) - U(t)\}|U(t), B(t)] = o(h). \qquad (24)$$

The transition probability density

$$p(\tau,u,v;t,x,y) = \frac{\partial}{\partial x}\frac{\partial}{\partial y} P[B(t) \le y, U(t) \le x|B(\tau) = u, U(\tau) = v] \qquad (25)$$

of the Markov process $(B(t), U(t))$ should therefore satisfy the forward equation

$$\frac{\partial p}{\partial t} = \beta\frac{\partial^2 p}{\partial x^2} + \beta\frac{\partial}{\partial x}(xp) - \frac{\partial}{\partial y}(xp) \qquad (26)$$

in $x$, $y$, $t$ and the corresponding backward equation in $v$, $u$, $\tau$. This is a two-dimensional diffusion equation that is singular since the second-order partial derivatives in $y$ do not appear.

# e. Pathological Jump Processes

In section b of this chapter, it was established that the transition probability function of a jump process satisfied both the backward and forward equations (b.3) and (b.4) under certain uniformity conditions. We shall construct simple examples of stationary denumerable state Markov chains with transition mechanism that do not satisfy the forward or backward equations. This "pathological" behavior is due to the fact that the boundedness of $p(t,x)$ assumed in condition (i) of section b is not satisfied. The first process constructed will behave very much like a pure birth process locally in time.

Let

$$b_{i,i} = -b_i$$
$$b_{i,i+1} = b_i > 0 \qquad \qquad (1)$$
$$b_{i,j} = 0 \text{ if } j \neq i, i+1$$

where $i, j = 0, 1, 2, \ldots$ . Here, of course, we are labeling the states of the process as $0, 1, 2, \ldots$ . The forward equations then are

$$p'_{i,j}(t) = b_{j-1}p_{i,j-1}(t) - b_j p_{i,j}(t) \text{ if } j \geq 1$$
$$p'_{i,0}(t) = -b_0 p_{i,0}(t), \text{ if } j = 0. \qquad \qquad (2)$$

Call the solution of this set of equations satisfying the condition

$$p_{i,j}(0) = \delta_{i,j} \qquad \qquad (3)$$

$p_{i,j}^{(0)}(t)$. It will be convenient to make use of the Laplace transform in the following discussion (see [1]). Let

$$\bar{p}_{i,j}^{(0)}(s) = \int_0^\infty e^{-st} p_{i,j}^{(0)}(t) \, dt. \qquad \qquad (4)$$

The differential equations (2) become the following simple set of linear equations in terms of the Laplace transforms $\bar{p}_{i,j}^{(0)}(s)$

$$s\bar{p}_{i,j}^{(0)}(s) - \delta_{i,j} = b_{j-1}\bar{p}_{i,j-1}^{(0)}(s) - b_j\bar{p}_{i,j}^{(0)}(s) \text{ if } j \geq 1$$
$$s\bar{p}_{i,0}^{(0)}(s) - \delta_{i,0} = -b_0\bar{p}_{i,0}^{(0)}(s) \qquad \qquad (5)$$

since

$$\int_0^\infty e^{-st} p'_{i,j}(t) \, dt = e^{-st} p_{i,j}(t) \Big|_0^\infty + s \int_0^\infty e^{-st} p_{i,j}(t) \, dt \qquad \qquad (6)$$
$$= -\delta_{i,j} + s\bar{p}_{i,j}^{(0)}(s).$$

A small computation leads to the solution

$$\bar{p}_{i,j}^{(0)}(s) = \begin{cases} 0 & \text{if } j < i \\ \prod_{k=i}^{j-1} b_k \Big/ \prod_{k=i}^{j} (s + b_k) & \text{if } j \geq i. \end{cases} \qquad \qquad (7)$$

The discussion of section b indicates that $p_{i,j}^{(0)}(t) \geq 0$

$$0 \leq \sum_j p_{i,j}^{(0)}(t) = 1 - h_i(t) \leq 1 \qquad \qquad (8)$$

and that $h_i(t)$ is nondecreasing. In fact, one can show that $h_i(t) \equiv 0$ if and only if

$$\sum 1/b_n = \infty \qquad \qquad (9)$$

and that otherwise $h_i(t)$ is strictly increasing (see Problem 10). Of
course, $\Sigma \, 1/b_n < \infty$ means that the $b_n$'s diverge rapidly as $n \to \infty$. The
specification (1) means that a system in state $i$ in a short time will with
large probability either remain in $i$ or else will move one step to the
right. The system cannot move to the left. The path of the system
through time can therefore be regarded as a continual drift to the right
through states of increasing magnitude. $h_i(t)$ can be thought of as the
probability that the system starting at $i$ has passed through all the
states in the finite time $t$. Further, we can provide a "sink" or absorbing
state at infinity where the system is lodged after it has passed through
all states. Thus $h_i(t)$ is the probability that the system has passed into
the adjoined state at infinity from state $i$ in time $t$. However, instead of
adjoining an absorbing state at infinity, we can immediately return the
system after it has drifted out of the finite states to the state $i$ with
probability $\alpha_i \geq 0$, $\Sigma \alpha_i = 1$, and start it out again with the same
transition mechanism as before. The probability $p_{i,j}^{(0)}(t)$ can then be
interpreted as the probability of going from $i$ to $j$ in time $t$ without
ever passing out of the set of finite states. The probability

$$p_{i,j}^{(1)}(t) = \int_0^t \sum_k \alpha_k p_{k,j}^{(0)}(t - \tau) \, dh_i(\tau) \tag{10}$$

is the probability of going from $i$ to $j$ in time $t$ while having passed out
of the set of finite states exactly once. Similarly

$$p_{i,j}^{(m+1)}(t) = \int_0^t \sum_k \alpha_k p_{k,j}^{(m)}(t - \tau) \, dh_i(\tau) \tag{11}$$

is the probability of going from $i$ to $j$ in time $t$ while having passed out
of the set of finite states exactly $m + 1$ times. Let

$$p_{i,j}(t) = \sum_{m=0}^{\infty} p_{i,j}^{(m)}(t). \tag{12}$$

By using Laplace transform techniques we shall show that

$$\sum_j p_{i,j}(t) \equiv 1. \tag{13}$$

Let $\bar{h}_i(s)$ and $\bar{p}_{i,j}^{(m)}(s)$ be the Laplace transforms of $h_i(t)$ and $p_{i,j}^{(m)}(t)$
respectively. Now (11) implies that

$$\bar{p}_{i,j}^{(m+1)}(s) = s \sum_k \alpha_k \bar{p}_{k,j}^{(m)}(s) \bar{h}_i(s) \tag{14}$$

so that

$$\bar{p}_{i,j}^{(m+1)}(s) = s^{m+1}(\sum_k \alpha_k \bar{h}_k(s))^m (\sum_k \alpha_k \bar{p}_{k,j}^{(0)}(s)) \bar{h}_i(s). \tag{15}$$

The transform $\bar{p}_{ij}(s)$ of $p_{ij}(t)$ is therefore given by

$$\bar{p}_{i,j}(s) = \bar{p}_{i,j}^{(0)}(s) + s\bar{h}_i(s)(\sum_k \alpha_k \bar{p}_{k,j}^{(0)}(s))/(1 - s \sum_k \alpha_k \bar{h}_k(s)). \tag{16}$$

Since

$$\bar{h}_i(s) = \frac{1}{s} - \sum_j \bar{p}_{i,j}^{(0)}(s) \tag{17}$$

it follows that

$$\sum_j \bar{p}_{i,j}(s) = \sum_j \bar{p}_{i,j}^{(0)}(s) + s\bar{h}_i(s)\left(\sum_j \sum_k \alpha_k \bar{p}_{k,j}^{(0)}(s)\right) \Big/ \left(1 - s \sum_k \alpha_k \bar{h}_k(s)\right)$$

$$= \frac{1}{s} - \bar{h}_i(s) + \bar{h}_i(s) = \frac{1}{s}, \tag{18}$$

the Laplace transform of one. Thus (13) is verified.

The backward equations are

$$p'_{i,j}(t) = b_i p_{i+1,j}(t) - b_i p_{i,j}(t) \tag{19}$$

or

$$s\bar{p}_{i,j}(s) - \delta_{i,j} = b_i \bar{p}_{i+1,j}(s) - b_i \bar{p}_{i,j}(s). \tag{20}$$

The probabilities $p_{i,j}^{(0)}(t)$ satisfy the forward equations since they were derived as solutions of those equations. They also satisfy the backward equations as is seen by simple inspection. Further, the Chapman-Kolmogorov equations follow also since

$$-\frac{d}{ds} \bar{p}_{i,j}^{(0)}(s) = \sum_k \bar{p}_{i,k}^{(0)}(s)\bar{p}_{k,j}^{(0)}(s). \tag{21}$$

However, $\sum_j p_{i,j}^{(0)}(t) < 1$ for $t > 0$ when $\sum 1/b_n < \infty$.

The probabilities $p_{i,j}(t)$ yield a true transition matrix since $\sum_j p_{i,j}(t) \equiv 1$. They satisfy the Chapman-Kolmogorov equation (see Problem VI.12). Moreover they behave locally like the probabilities $p_{i,j}^{(0)}(t)$ since the derivatives

$$p'_{i,j}(0) = \begin{cases} -b_i & \text{if } j = i \\ b_i & \text{if } j = i + 1 \\ 0 & \text{otherwise.} \end{cases} \tag{22}$$

This follows from the fact that $\lim_{s \to \infty} [s^2\bar{p}_{i,j}(s) - s\delta_{i,j}] = p'_{i,j}(0)$. Simple substitution in (20) indicates that the probabilities satisfy the backward equations. However, they do not satisfy the forward equations. Instead forward inequalities

$$p'_{i,j}(t) > b_{j-1}p_{i,j-1}(t) - b_j p_{i,j}(t), j \geq 1$$
$$p'_{i,0}(t) > -b_0 p_{i,0}(t) \tag{23}$$

are obtained for $t > 0$. Limiting stationary probabilities $p_j = \lim_{t \to \infty} p_{i,j}(t)$ can be evaluated

$$p_j = \lim_{s \to 0} s\bar{p}_{i,j}(s) = \frac{1}{b_j} \sum_{k \leq j} \alpha_k \Bigg/ \left( \sum_k \alpha_k \sum_{m \geq k} \frac{1}{b_m} \right). \tag{24}$$

Let us take these limiting probabilities as the initial distribution so that we have a stationary Markov chain. If time is reversed a new stationary Markov chain with the same stationary initial distribution $p_i$ and transition probabilities

$$q_{i,j}(t) = p_j p_{j,i}(t)/p_i \tag{25}$$

is obtained. The backward and forward equations of the new chain are the forward and backward equations respectively of the original chain. Thus, the transition probabilities $q_{i,j}(t)$ of the new chain now satisfy the forward equations but only backward inequalities are satisfied.

By combining the chains with transition probabilities $p_{i,j}(t)$ and $q_{i,j}(t)$ respectively in the proper manner a chain whose transition probabilities satisfy neither backward nor forward equations can be constructed. Let us consider a chain with doubly indexed states $(i,i')$, $i, i' = 0, 1, 2, \ldots$ . Let the stationary probability of being in the state $(i,i')$ be given by

$$P[X(t) = (i,i')] = p_i p_{i'}. \tag{26}$$

Further set the transition probability equal to

$$P[X(t) = (j,j')|X(\tau) = (i,i')]$$
$$= p_{i,j}(t - \tau)q_{i',j'}(t - \tau), \tau < t. \tag{27}$$

The transition probabilities of this chain satisfy neither the backward nor forward equations.

## f. Problems

1. Let $\mathbf{M}_1$, $\mathbf{M}_2$, . . . be independent and identically distributed random matrices. Set $\mathbf{T}_n = \prod_{j=1}^{n} \mathbf{M}_j$. Show that the sequence of random matrices $\mathbf{T}_n$ can be considered a Markov process. Examine $E\mathbf{T}_n$ in terms of the distribution of the random matrices $\mathbf{M}_k$. Consider also $E\mathbf{T}_n^k$.

2. Suppose the $\mathbf{M}_k$ in the previous example are random transition probability matrices. Can you give more detailed information about the limit behavior of $\mathbf{T}_n$ as $n \to \infty$?

3. Consider a strictly stationary process $X_n$ $n = 0$, $\pm 1$, $\pm 2$, . . . where the random variables $X_n$ take only the two possible values $0$, $1$. Let $Y_n$ $n = 0$, $\pm 1$, . . . be the stationary process obtained by setting $Y_n$ equal to the binary expansion $Y_n = . X_n X_{n-1} \cdots$. Show that $Y_n$ is Markovian and determine the transition mechanism of this Markov process.

4. Consider the following idealization of a telephone trunking problem. Suppose infinitely many trunks or channels are available and that the probability of a telephone conversation ending in the interval $(t, t + h)$ is $\mu h$ to the first order. Further assume the probability of a new call coming in during the interval $(t, t + h)$ is $\lambda h$ to the first order. The system is in state $n$ if $n$ lines are busy. Show that the generating function $P(s,t) = \sum_n p_{0,n}(t) s^n$ satisfies

$$\partial P/\partial t = (1 - s)\left\{-\lambda P + \mu \frac{\partial P}{\partial s}\right\}.$$

Solve for $P(s,t)$ and find the limit behavior of $p_{0,n}(t)$ as $t \to \infty$.

5. Let $X(t)$ be the Wiener process. Consider the derived process $Y(t) = X(t)$ if $\max_{0 \le \tau \le t} X(\tau) < a$ and $a$ otherwise where $a > 0$. Is $Y(t)$ Markovian? Find the probability structure of the process $Y(t)$. Can you physically interpret what has been done to the process $X(t)$ to obtain the process $Y(t)$? Show that the transition probability density of the $Y$ process satisfies the heat equation away from $a$?

6. Let $X(t)$ be the Wiener process as in the previous example. Let

$$Y(t) = \begin{cases} X(t) & \text{if } X(t) \le a \\ 2a - X(t) & \text{if } X(t) \ge a. \end{cases}$$

Answer the questions posed in the previous problem with respect to this process.

7. Let $X(t)$ be the Wiener process. Let $Y(t) = X(t) - [X(t)]$ where $[x]$ is the greatest integer less than $x$. Answer the questions posed in the previous problem with respect to the process $Y(t)$.

8. Under the assumption that $a(\tau,\xi)$ is bounded away from zero and infinity and differentiable in $\xi$, show that the differential equation $\dfrac{\partial F}{\partial \tau} + \dfrac{1}{2} a(\tau,\xi) \dfrac{\partial^2 F}{\partial \xi^2} + b(\tau,\xi) \dfrac{\partial F}{\partial \xi} = 0$ can be reduced to an equation of the form $\dfrac{\partial G}{\partial \tau} + \dfrac{1}{2} \dfrac{\partial^2 G}{\partial \eta^2} + c(\tau,\eta) \dfrac{\partial G}{\partial \eta} = 0$ by the change of variable $\eta = \int_0^\xi dy/\sqrt{a(\tau,y)}$.

9. Show that if $g(t,x)$ satisfies a Hölder condition of the type (c.18) and

$$G(\tau,\xi) = \int_t^T dt \int_{-\infty}^\infty g(t,x)u(\tau,\xi;t,x)\, dx$$

that $\dfrac{\partial G}{\partial \tau}, \dfrac{\partial G}{\partial \xi}, \dfrac{\partial^2 G}{\partial \xi^2}$ exist and $G$ satisfies the differential equation

$$\frac{\partial G}{\partial \tau} + \frac{1}{2} \frac{\partial^2 G}{\partial \xi^2} = -g(\tau,\xi).$$

Hint: See the paper of E. E. Levi [49].

10. Show that in the case of a pure birth process $\sum_j p_{i,j}^{(0)}(t) \equiv 1$ if and only if $\sum 1/b_n = \infty$.

11. Show that the $p_{i,j}^{(0)}(t)$ of section e satisfy the backward equations.

12. Show that the $p_{i,j}(t)$ of section e satisfy the Chapman-Kolmogorov equation.

# Notes

1. Generalizations of the Markovian concept for processes with multidimensional time were initially discussed by P. Lévy (see [51]). More recently there has been renewed interest on these questions and some rather interesting results (see the papers of Dobrushin [A3] and L. Pitt [A12]).

2. The discussion of Markov processes given in this chapter is basically classical and analytic in character. This was thought to be a most effective presentation in an introductory book. Of course, this means that much of the recent work on Markov processes, particularly that concerned with analysis of regularity properties of

sample functions of the process as well as that utilizing semigroup techniques to obtain properties of the process, is not discussed. We make a few brief remarks to indicate how a semigroup arises naturally when dealing with a stationary transition mechanism. It is obvious that one is dealing with a semigroup of matrices in the case of a Markov chain. Nonetheless, it is worthwhile making a few remarks in the case of a general state space. Let $P(t,x,A)$ be a stationary transition probability function

$$P(t,x,A) = P[X(t + \tau)\epsilon A | X(\tau) = x], \, t \geq 0,$$

satisfying the analogues of conditions (i)–(iii) of section a for a stationary transition mechanism. We can alternatively construct a semigroup of operators acting on probability measures or functions in terms of $P(t,x,A)$. First consider a probability measure $\mu$ on $\mathfrak{F}_X$. The transition probability function takes $\mu$ into a probability measure $\nu$

$$\nu(A) = \int_{\Omega_X} \mu(dx)P(t,x,A) = (T^{(t)}\mu)(A).$$

The family of operators $T^{(t)}$, $t \geq 0$, is a semigroup

$$T^{(t)}T^{(\tau)} = T^{(t+\tau)}, \, t, \tau \geq 0,$$

since

$$\int_{\Omega_X} P(t,x,dy)P(\tau,y,A) = P(t + \tau, x, A).$$

One can also construct a family of operators $S^{(t)}$, $t \geq 0$, taking bounded functions $f$ into bounded functions $g$

$$g(x) = \int_{\Omega_X} f(y)P(t,x,dy) = (S^{(t)}f)(x).$$

Notice that $S^{(t)}$ is a positive operator since it takes non-negative functions into non-negative functions. The family of operators $S^{(t)}$, $t \geq 0$, is a semigroup

$$S^{(t)}S^{(\tau)} = S^{(t+\tau)}, \, t, \tau \geq 0,$$

because of (a.1).

3. Kolmogorov's paper on analytic methods in dealing with jump and continuous Markov processes [43] is an early basic paper. It is well worth reading since many of the approaches later elaborated and refined are presented there. Under appropriate conditions one can show that there is a representation of diffusion processes with continuous sample functions (see Doob [12]). Even when dealing with jump processes one often wants a representation with right continuous sample functions. The example mentioned in the last paragraph of section e shows that this may not be possible to achieve.

# VII

## WEAKLY STATIONARY PROCESSES AND RANDOM HARMONIC ANALYSIS

### a. Definition

In Chapter V stationary processes (sometimes called strictly stationary processes) were motivated and discussed as models of natural phenomena. We will now consider a related class of processes commonly termed *weakly stationary* processes. Though the name would appear to imply a larger class than that of the stationary processes, this is not quite the case. These processes are characterized essentially by their second moment properties. Let $X_t(w)$, $-\infty < t < \infty$, be a continuous time parameter complex-valued process with finite second moments $E|X_t(w)|^2 < \infty$. For convenience we shall take its mean $EX_t(w) \equiv 0$. $X_t(w)$ is called a weakly stationary process if its covariance function

$$EX_t(w)\bar{X}_\tau(w) = r(t,\tau) = r(t - \tau) \tag{1}$$

depends only on the time difference $t - \tau$ and is continuous. Notice that the continuity of $r(t)$ implies that $X_t(w)$ is continuous in the mean, that is,

$$E|X_t(w) - X_\tau(w)|^2 = 2r(0) - r(t - \tau) - r(\tau - t) \to 0 \tag{2}$$

as $t - \tau \to 0$. It is clear that a stationary process with finite second moments must be weakly stationary while the converse is not true. On the other hand, if a stationary process has infinite second moments it is meaningless to speak of weak stationarity.

The covariance function $r(t)$ is positive definite (see section b, Chapter IV) since for any given finite number of points $t_1, \ldots, t_k$ and complex numbers $c_1, \ldots, c_k$

$$\sum_{i,j=1}^{k} c_i\bar{c}_j r(t_i - t_j) = E|\sum_{i=1}^{k} c_i X_{t_i}(w)|^2 \geq 0. \tag{3}$$

Now Bochner's theorem (see section b of Chapter VI) implies that

$r(t)$ is, except for a constant factor, a characteristic function. Therefore $r(t)$ has a Fourier-Stieltjes representation

$$r(t) = \int_{-\infty}^{\infty} e^{it\lambda} \, dF(\lambda) \tag{4}$$

where $F(\lambda)$ is a nondecreasing function with $F(+\infty) - F(-\infty) = \lim_{L \to \infty} [F(L) - F(-L)] = r(0)$ finite (not necessarily one). For convenience $F(\lambda)$ is assumed to be continuous to the right. The function $F(\lambda)$ is called the *spectral distribution function* of the process $X_t(w)$. Here $dF(\lambda)$ is the weight to be attributed to $e^{it\lambda}$ in the frequency resolution of $r(t)$. We refer to (4) as a frequency resolution because the covariances $r(t)$ have been partitioned into components $e^{it\lambda} \, dF(\lambda)$ corresponding to the frequencies $\lambda$. Notice that in the case of a real-valued process $X_t$ $r(t) = r(-t)$ so that the spectral mass is symmetrically located about zero, that is, $dF(\lambda) = dF(-\lambda)$. If $F$ is absolutely continuous with respect to Lebesgue measure, its derivative $f(\lambda) = F'(\lambda)$ is called the *spectral density function* of the process.

Suppose the continuous parameter process $X_t$ is observed only at the discrete time points $t = k\Delta$, $k = 0, \pm 1, \ldots$ . The covariance function of the corresponding discrete parameter process $X_{k\Delta}$ is $r(k\Delta) = EX_{(j+k)\Delta}\bar{X}_{j\Delta}$ (also called weakly stationary because the covariance function depends only on time differences). Notice that $r(k\Delta)$, $k = 0, \pm 1, \ldots$ is a *positive definite sequence*. Now

$$r(k\Delta) = \int_{-\infty}^{\infty} e^{ik\Delta\lambda} \, dF(\lambda)$$

$$= \sum_{j=-\infty}^{\infty} \int_{\frac{(2j-1)\pi}{\Delta}}^{\frac{(2j+1)\pi}{\Delta}} e^{ik\Delta\lambda} \, dF(\lambda)$$

$$= \int_{-\frac{\pi}{\Delta}}^{\frac{\pi}{\Delta}} e^{ik\Delta\lambda} \, dG(\lambda) \tag{5}$$

where

$$G(\lambda) = \sum_{j=-\infty}^{\infty} \left\{ F\left(\lambda + \frac{2j\pi}{\Delta}\right) - F\left(\frac{(2j-1)\pi}{\Delta}\right) \right\}.$$

The range of the frequency resolution of the sequence $r(k\Delta)$ is $\left[-\frac{\pi}{\Delta}, \frac{\pi}{\Delta}\right]$, a finite range. The spectral distribution function $G(\lambda)$ of the discrete parameter process $X_{k\Delta}$ is obtained from the spectral distribution function $F(\lambda)$ of the continuous parameter process by a folding operation (see (5)), an effect called "aliasing." A point $\lambda$ is called a point of increase of $F$ if for all $\lambda_1, \lambda_2$ with $\lambda_1 < \lambda < \lambda_2$ one has $F(\lambda_2) - F(\lambda_1) > 0$. Notice that aliasing means that the spectrum of the continuous parameter process $X_t$ cannot be completely determined from that of the process discretely sampled $X_{k\Delta}$ unless the spectrum of $X_t$ is "band limited," that is, the points of increase of $F$ are limited to a finite interval. Nonetheless, one can, of course, obtain considerable information about the spectrum through discrete sampling. The spectral density $g(\lambda)$ of $X_{k\Delta}$ is given in terms of $f(\lambda)$ by

$$g(\lambda) = \sum_{j=-\infty}^{\infty} f\left(\lambda + \frac{2j\pi}{\Delta}\right), \quad |\lambda| < \frac{\pi}{\Delta}. \tag{6}$$

As is implied by (5), a formula paralleling (4) holds for discrete parameter weakly stationary processes. Let $X_k$, $k = 0, \pm 1, \ldots$, $EX_k \equiv 0$, be a discrete parameter stationary process so that

$$EX_k \bar{X}_j = r_{k-j}. \tag{7}$$

The covariance sequence $r_k$ is positive definite and hence by an antecedent of Bochner's theorem due to Herglotz (see [53]) $r_k$ has the representation

$$r_k = \int_{-\pi}^{\pi} e^{ik\lambda} \, dF(\lambda) \tag{8}$$

with $F$ a nondecreasing bounded function. The function $F$ as before is called the spectral distribution function of $X_k$.

A simple example of a discrete time parameter stationary process is given by

$$X_t = \sum_{k=1}^{n} c_k e^{i(t\lambda_k + \varphi_k)}, \quad t = 0, \pm 1, \ldots, \tag{9}$$

where the $c_k$ are real constants, the $\lambda_k$ real numbers in $[-\pi, \pi]$ and the $\varphi_k$ independent random variables uniformly distributed on $[-\pi, \pi]$. This example arises in the theory of noise (see [26]), where the noise current at time $t$ is considered a superposition of alternating current components of frequency $\lambda_k/2\pi$ cycles per second with amplitudes $c_k$ and random phases $\varphi_k$. It is called the model of random phases. The mean value of the process is zero

$$EX_t = \sum_{k=1}^{n} \frac{1}{2\pi} c_k \int_{-\pi}^{\pi} e^{i(t\lambda_k + \varphi_k)} \, d\varphi_k \equiv 0 \tag{10}$$

and the covariance function

$$EX_t \bar{X}_\tau = \sum_{k=1}^{n} c_k^2 \frac{1}{2\pi} \int_{-\pi}^{\pi} e^{i(t-\tau)\lambda_k} d\varphi_k$$

$$= \sum_{k=1}^{n} c_k^2 e^{i(t-\tau)\lambda_k} = \int_{-\pi}^{\pi} e^{i(t-\tau)\lambda} dF(\lambda) \tag{11}$$

because of the independence of the $\varphi_k$. The spectral distribution function $F$ is the nondecreasing step function

$$F(\lambda) = \sum_{\lambda_k \leq \lambda} c_k^2. \tag{12}$$

The second model is a representative example of some that enter in econometric discussions [48]. Let $X_t$, $Y_t$ be the price and supply respectively of a specific commodity at time $t = 0, \pm 1, \ldots$. The price $X_t$ and the supply $Y_t$ are linked by the difference equations

$$X_t = \alpha - \beta Y_t + \eta_t'$$
$$Y_t = \gamma + \delta X_{t-1} + \eta_t'' \tag{13}$$

where $\alpha$, $\beta$, $\gamma$, $\delta$ are real constants and $\eta_t'$, $\eta_t''$ are random variables representing the random disturbances that this aspect of the economic system are exposed to. On solving for $X_t$ we obtain

$$X_t = \alpha - \beta\gamma - \beta\,\delta X_{t-1} + \eta_t' - \beta\eta_t''. \tag{14}$$

This is a stochastic difference equation of order 1. Difference equations of this type will be discussed in some detail in section c.

It is clear from the representations (4) and (8) that the second-order properties of weakly stationary processes can be examined using either their covariances or spectra since they are equivalent. In certain communication problems in engineering, the message transmitted is represented by a continuous time parameter weakly stationary process $X_t$. Suppose the message is passed through a linear filter $\mathcal{F}$. Then the output of the filter is given by

$$Y_t = \mathcal{F}X_t = \int_{-\infty}^{\infty} g(t - \tau)X(\tau)\, d\tau$$

$$= \int_{-\infty}^{t} g(t - \tau)X(\tau)\, d\tau. \tag{15}$$

It is physically plausible to assume $g(t) = 0$ for $t < 0$ and we have done this. The function $g(t)$ is called the *transient response function* because

$g(t) \, dt$ is the output at time $t$ due to a small pulse of height one and length $dt$ at $t = 0$. To make (15) meaningful we assume $g$ to be integrable. Consider the covariance function of the output

$$
\begin{aligned}
\operatorname{cov}(Y_t, Y_\tau) &= \iint g(t - \alpha)g(\tau - \beta) \operatorname{cov}(X_\alpha, X_\beta) \, d\alpha \, d\beta \\
&= \iint g(t - \alpha)g(\tau - \beta)r(\alpha - \beta) \, d\alpha \, d\beta \qquad (16) \\
&= \int r(u) \int g(\tau - t + u - \alpha)g(-\alpha) \, d\alpha \, du.
\end{aligned}
$$

Notice that the covariance function of the output is obtained from the covariance function of the input by an operation that is computationally often quite messy, the convolution operation. The corresponding transformation in the spectral domain is much simpler since as we shall see it is simply a multiplication. Using the representation of the covariance function in terms of the spectral distribution function

$$
\begin{aligned}
\operatorname{cov}(Y_t, Y_\tau) &= \iint g(t - \alpha)g(\tau - \beta) \int e^{i(\alpha - \beta)\lambda} \, dF(\lambda) \, d\alpha \, d\beta \\
&= \int e^{i(t - \tau)\lambda} |\gamma(\lambda)|^2 \, dF(\lambda) \qquad (17)
\end{aligned}
$$

where

$$
\gamma(\lambda) = \int e^{-it\lambda} g(t) \, dt. \qquad (18)
$$

Thus the spectral distribution function $G(\lambda)$ of the output is related to that of the input by the simple relation

$$
dG(\lambda) = |\gamma(\lambda)|^2 \, dF(\lambda). \qquad (19)
$$

The function $\gamma(\lambda)$ is called the frequency response function because its absolute value measures the amplification at frequency $\lambda$ due to the filter. The preference for speaking in spectral terms in the engineering literature is due in part to the simple transformation properties referred to above.

# b. Harmonic Representation of a Stationary Process and Random Integrals

A stationary process $X_t$ has a random Fourier representation paralleling that of its covariance function (see (a.4) and (a.8)). A detailed derivation of this representation will be carried out in the case of a continuous time parameter process. It is readily seen that an analogous argument will yield the corresponding result if time is discrete.

Let $F(\lambda)$ be the spectral distribution function of $X_t$. Let $\lambda, \mu, \lambda < \mu,$

be continuity points of $F$. We introduce the random variable

$$Z_T(\mu,\lambda) = \frac{1}{2\pi} \int_{-T}^{T} X_t \int_{\lambda}^{\mu} e^{-it\alpha} \, d\alpha \, dt \tag{1}$$

$$= \frac{1}{2\pi} \int_{-T}^{T} X_t \frac{e^{-it\mu} - e^{-it\lambda}}{-it} \, dt.$$

Now if $0 < \tau \le T$

$$E|Z_T(\mu,\lambda) - Z_\tau(\mu,\lambda)|^2 = \frac{1}{4\pi^2} E \left| \int_{\tau \le |t| < T} X_t \int_{\lambda}^{\mu} e^{-it\alpha} \, d\alpha \, dt \right|^2$$

$$= \frac{1}{4\pi^2} \int_{-\infty}^{\infty} dF(u) \left| \int_{\tau \le |t| < T} \int_{\lambda}^{\mu} e^{-it(\alpha - u)} \, d\alpha \, dt \right|^2 \tag{2}$$

making use of the Fourier representation of the covariance function. The function

$$\int_{\tau \le |t| < T} \int_{\lambda}^{\mu} e^{-it(\alpha - u)} \, d\alpha \, dt \tag{3}$$

converges uniformly to zero as $\tau, T \to \infty$ if $|u-\mu|$, $|u-\lambda| > \varepsilon$ where $\varepsilon$ is any fixed positive number. For

$$\int_{\tau}^{T} \int_{\lambda}^{\mu} \cos t(\alpha - u) \, d\alpha \, dt = \int_{\tau}^{T} \frac{\sin t(\mu - u) - \sin t(\lambda - u)}{t} \, dt \tag{4}$$

and

$$\left| \int_{\tau}^{T} \frac{\sin t(\mu - u)}{t} \, dt \right| \le \left| \frac{\cos t(\mu - u)}{t(\mu - u)} \right|_{\tau}^{T} \right| + \left| \int_{\tau}^{T} \frac{\cos t(\mu - u)}{t^2(\mu - u)} \, dt \right|$$

$$\le 2 \left( \frac{1}{\tau} + \frac{1}{T} \right) \frac{1}{|\mu - u|}.$$

Also a direct estimate using the oscillatory character of $\sin t/t$ shows that expression (3) is uniformly bounded in absolute value. Thus

$$E|Z_T(\mu,\lambda) - Z_\tau(\mu,\lambda)|^2 \to 0 \tag{5}$$

as $\tau, T \to \infty$ and hence, by the Riesz-Fischer theorem (see section a of Chapter IV) $Z_T(\mu,\lambda)$ converges in mean square as $T \to \infty$ to a limiting random variable that we shall call $Z(\mu,\lambda)$. We consider now the computation of the covariance

$$E(Z(\mu,\lambda)\overline{Z(\mu',\lambda')}) \tag{6}$$

where, for convenience, $\mu$, $\lambda$, $\mu'$, $\lambda'$ are taken to be points of continuity of the spectral distribution function $F(\lambda)$ of the process $X_t$. These covariances exist since

$$E|Z(\mu,\lambda)|^2 = \lim_{T \to \infty} E|Z_T(\mu,\lambda)|^2 \tag{7}$$

is finite. Now

$$E(Z(\mu,\lambda)\overline{Z(\mu',\lambda')}) = \lim_{T\to\infty} E(Z_T(\mu,\lambda)\overline{Z_T(\mu',\lambda')})$$

$$= \lim_{T\to\infty} \frac{1}{4\pi^2} \int_{-\infty}^{\infty} dF(u) \int_0^T 2 \int_\lambda^\mu \cos t(\alpha - u)\, d\alpha\, dt$$

$$\int_0^T 2 \int_{\lambda'}^{\mu'} \cos t'(\alpha' - u)\, d\alpha'\, dt'. \quad (8)$$

Note that

$$\int_0^T \int_\lambda^\mu \cos t(\alpha - u)\, d\alpha\, dt = \int_0^T \frac{\sin t(\mu - u) - \sin t(\lambda - u)}{t}\, dt. \quad (9)$$

We now use the fact that

$$\lim_{T\to\infty} \int_0^T \frac{\sin ta}{t}\, dt = \int_0^\infty \frac{\sin ta}{t}\, dt = \frac{\pi}{2} \quad (10)$$

for $a > 0$ (see Courant's *Integral and Differential Calculus*, p. 450). Thus

$$\lim_{T\to\infty} \int_0^T \int_\lambda^\mu \cos t(\alpha - u)\, d\alpha\, dt = \begin{cases} \pi \text{ if } \lambda < u < \mu \\ 0 \text{ if } \mu < u \text{ or } u < \lambda \\ \dfrac{\pi}{2} \text{ if } u = \lambda \text{ or } u = \mu. \end{cases} \quad (11)$$

In fact, the convergence is uniform if $u$ is bounded away from $\lambda$ and $\mu$. On interchanging limiting operations in (8) it is seen that

$$E(Z(\mu,\lambda)\overline{Z(\mu',\lambda')}) = \begin{cases} \displaystyle\int_{\max(\lambda,\lambda')}^{\min(\mu,\mu')} dF(u) = F(\min(\mu,\mu')) - F(\max(\lambda,\lambda')) \\ \qquad\qquad \text{if } \min(\mu,\mu') > \max(\lambda,\lambda') \quad (12) \\ 0 \text{ otherwise.} \end{cases}$$

Thus, if $(\lambda,\mu)$ and $(\lambda',\mu')$ are disjoint intervals $Z(\mu,\lambda)$ and $Z(\mu',\lambda')$ are orthogonal, that is, they are uncorrelated. Notice that (12) implies that

$$E|Z(\mu,\lambda)|^2 = F(\mu) - F(\lambda). \quad (13)$$

Since

$$E|Z(\mu,\lambda) - Z(\mu,\lambda')|^2 = |F(\lambda) - F(\lambda')| \to 0 \quad (14)$$

as $\lambda, \lambda' \to -\infty$

$$\lim_{\lambda\to-\infty} Z(\mu,\lambda) = Z(\mu) \quad (15)$$

exists as a limit in mean square. Using the definition (15) it is readily seen that

$$Z(\mu,\lambda) = Z(\mu) - Z(\lambda). \quad (16)$$

Property (12) can be rewritten in terms of the process $Z(\mu)$ as

$$E(Z(\mu) - Z(\lambda))(\overline{Z(\mu') - Z(\lambda')}) = 0 \tag{17}$$

if $(\lambda,\mu)$, $(\lambda',\mu')$ are disjoint and

$$E|Z(\mu) - Z(\lambda)|^2 = F(\mu) - F(\lambda), \tag{18}$$

or even more intuitively in differential notation

$$E \, dZ(\lambda) \, \overline{dZ(\mu)} = \delta_{\lambda,\mu} \, dF(\lambda) \tag{19}$$

where $\delta_{\lambda,\mu}$ is the Kronecker delta. Such processes $Z(\lambda)$ are typically called *processes with orthogonal increments*. We have in the discussion above limited ourselves to continuity points $\lambda$ of $F(\lambda)$ to avoid the additional fuss and notation required at discontinuity points. However, the process $Z(\lambda)$ is readily defined at discontinuity points $\lambda$ of $F$ by setting

$$Z(\lambda) = \lim_{\mu \to \lambda+} Z(\mu) \tag{20}$$

where $\mu$ approaches $\lambda$ from above through continuity points of $F$. Of course, this can be done since the continuity points of $F$ are dense everywhere.

A random Stieltjes integral of the form

$$\int g(\lambda) \, dZ(\lambda) \tag{21}$$

with $g$ a fixed function and $Z$ a process of orthogonal increments can be introduced in the following way. The integral is first defined for step functions, that is, functions $g$ of the form

$$g(\lambda) = \begin{cases} g_k & \text{if } a_{k-1} < \lambda \leq a_k \quad k = 1, \, \ldots \, , i \\ 0 & \text{otherwise} \end{cases} \tag{22}$$

where $-\infty < a_0 < \cdots < a_i < \infty$. For such a function $g$ we define $\int g(\lambda) \, dZ(\lambda)$ as

$$\int g(\lambda) \, dZ(\lambda) = \sum_{k=1}^{n} g_k[Z(a_k) - Z(a_{k-1})]. \tag{23}$$

Notice that

$$E\left| \int g(\lambda) \, dZ(\lambda) \right|^2 = \sum_{k=1}^{n} |g_k|^2[F(a_k) - F(a_{k-1})]$$

$$= \int |g(\lambda)|^2 \, d\mu(\lambda) \tag{24}$$

since $Z$ is a process with orthogonal increments and $\mu$ *is the measure on the real line generated by F.* The integral will now be defined for any bounded continuous function $g$. Given any such continuous function $g$, there is a sequence of step functions $g_n$ of the form (22) such that

$$\int |g(\lambda) - g_n(\lambda)|^2 \, d\mu \to 0 \tag{25}$$

as $n \to \infty$. Let $J_n$ be the random variable

$$J_n = \int g_n(\lambda) \, dZ(\lambda). \tag{26}$$

Then

$$E|J_n - J_m|^2 = \int |g_n(\lambda) - g_m(\lambda)|^2 \, d\mu \to 0 \tag{27}$$

as $n, m \to \infty$ by (24) and (25). By the Riesz-Fischer theorem $J_n$ converges in the mean square to a random variable $J$ with

$$E|J|^2 = \lim_{n \to \infty}{}' E|J_n|^2 = \int |g(\lambda)|^2 \, d\mu(\lambda). \tag{28}$$

It is natural to define $\int g(\lambda) \, dZ(\lambda)$ as $J$ since $J$ does not depend on the particular sequence $g_n(\lambda)$ by which $g$ is approximated. This follows from a simple application of the Minkowski inequality (see [1]). Actually this integral can be defined for the larger class of Borel functions $g$ for which $\int |g(\lambda)|^2 \, d\mu(\lambda)$ is finite by a similar argument. This integral has the following typical properties of an integral and these properties can be derived formally by considering approximating step functions and going to the limit:

1. $\int [ag(\lambda) + bh(\lambda)] \, dZ(\lambda) = a\int g(\lambda) \, dZ(\lambda) + b\int h(\lambda) \, dZ(\lambda)$  (29)

2. $\lim\limits_{n \to \infty} \int g_n(\lambda) \, dZ(\lambda) = \int g(\lambda) \, dZ(\lambda)$  (30)

if and only if

$$\int |g_n(\lambda) - g(\lambda)|^2 \, d\mu(\lambda) \to 0 \tag{31}$$

as $n \to \infty$.

3. $E\int g(\lambda) \, dZ(\lambda) \overline{\int h(\lambda) \, dZ(\lambda)} = \int g(\lambda)\overline{h(\lambda)} \, d\mu(\lambda).$  (32)

We shall now show that *a weakly stationary process $X_t$ has a random Fourier representation in terms of a process $Z(\lambda)$ with orthogonal increments,* that is,

$$X_t = \int_{-\infty}^{\infty} e^{it\lambda} \, dZ(\lambda), \tag{33}$$

where $Z(\lambda)$ is the process given by (15). Further

$$E \, dZ(\lambda) \, \overline{dZ(\mu)} = \delta_{\lambda,\mu} \, dF(\lambda) \tag{34}$$

*where $F(\lambda)$ is the spectral distribution function of $X_t$.* Let $\lambda$, $\mu$ be continuity points of $F(\lambda)$, $\lambda < \mu$. Now

$$
\begin{aligned}
EX_t\overline{Z_T(\mu,\lambda)} &= \frac{1}{2\pi} \int_{-T}^{T} r(t - \tau) \int_{\lambda}^{\mu} e^{i\tau\alpha} \, d\alpha \, d\tau \\
&= \int_{-\infty}^{\infty} e^{itu} \, dF(u) \frac{1}{\pi} \int_{0}^{T} \int_{\lambda}^{\mu} \cos \tau(\alpha - u) \, d\alpha \, d\tau \\
&= \int_{-\infty}^{\infty} e^{itu} \, dF(u) \frac{1}{\pi} \int_{0}^{T} \frac{\sin \tau(\mu - u) - \sin \tau(\lambda - u)}{\tau} \, d\tau \\
&\longrightarrow \int_{\lambda}^{\mu} e^{itu} \, dF(u)
\end{aligned}
\tag{35}
$$

as $T \to \infty$, making use of (1), (10), and the Fourier representation of the covariance function $r(t)$. But

$$
EX_t\overline{[Z(\mu) - Z(\lambda)]} = EX_t \overline{\int_{\lambda}^{\mu} dZ(\lambda)} = \int_{\lambda}^{\mu} e^{itu} \, dF(u)
\tag{36}
$$

since $Z(\mu) - Z(\lambda)$ is the limit in mean square of $Z_T(\lambda,\mu)$ as $T \to \infty$. There is a sequence $g_n(\lambda)$ of step functions of the form (22) with jumps at continuity points of $F(\lambda)$ such that

$$
\int |g_n(\lambda) - e^{it\lambda}|^2 \, dF(\lambda) \to 0
\tag{37}
$$

as $n \to \infty$. Therefore $\int g_n(\lambda) \, dZ(\lambda) \to \int e^{it\lambda} \, dZ(\lambda)$ as $n \to \infty$. Relation (36) implies that

$$
EX_t\overline{\int g_n(\lambda) \, dZ(\lambda)} = \int e^{it\lambda}\overline{g_n(\lambda)} \, d\mu(\lambda)
\tag{38}
$$

and hence

$$
\begin{aligned}
EX_t\overline{\int e^{it\lambda} \, dZ(\lambda)} &= \lim_{n \to \infty} EX_t\overline{\int g_n(\lambda) \, dZ(\lambda)} \\
&= \lim_{n \to \infty} \int e^{it\lambda}\overline{g_n(\lambda)} \, d\mu(\lambda) = \int dF(\lambda).
\end{aligned}
\tag{39}
$$

Now

$$
E|X_t - \int e^{it\lambda} \, dZ(\lambda)|^2 = E|X_t|^2 - 2 \operatorname{Re} EX_t\overline{\int e^{it\lambda} \, dZ(\lambda)} + E|\int e^{it\lambda} \, dZ(\lambda)|^2
$$
$$
= 0
\tag{40}
$$

using (a.4), (32), and (39). But this implies that

$$
X_t = \int_{-\infty}^{\infty} e^{it\lambda} \, dZ(\lambda)
\tag{41}
$$

with probability one.

Of course, a corresponding representation holds in the case of a discrete time parameter weakly stationary process $X_n$, $n = 0, \pm 1,$ $\pm 2, \ldots$ , except that in this case the range of integration in the random Fourier representation is $[-\pi, \pi]$ instead of $(-\infty, \infty)$

$$X_n = \int_{-\pi}^{\pi} e^{in\lambda}\, dZ(\lambda), \; E\, dZ(\lambda)\overline{dZ(\mu)} = \delta_{\lambda,\mu}\, dF(\lambda). \tag{42}$$

Here, $F(\lambda)$ is as before the spectral distribution function of the process $X_n$. Notice that in the representations (41) and (42) there is a random frequency resolution of the processes with $dZ(\lambda)$ the random amplitude of $e^{it\lambda}$. The random amplitudes $dZ(\lambda)$, $dZ(\mu)$ corresponding to distinct frequencies $\lambda \neq \mu$ are orthogonal and the variance of $dZ(\lambda)$ is given by $dF(\lambda)$, (see (42)).

We shall use the random spectral representation of a weakly stationary process obtained above to derive a mean square ergodic theorem for a discrete parameter stationary process. Let $X_n$ be a weakly stationary process with mean $EX_n = m$ not necessarily zero. Our object is to investigate the behavior of the time average

$$\frac{1}{n}\sum_{k=1}^{n} X_k \tag{43}$$

as $n \to \infty$. Let $\Delta Z(0) = Z(0) - \lim_{\varepsilon \to 0+} Z(-\varepsilon) = Z(0) - Z(0-)$ be the jump of $Z(\lambda)$ at zero. Of course $\Delta Z(0) = 0$ if there is no jump. We shall show that *the time average* $\dfrac{1}{n}\sum_{k=1}^{n} X_k$ *converges to* $\Delta Z(0) + m$ *as* $n \to \infty$. The random variable $X_k - m$ has the Fourier representation

$$X_k - m = \int_{-\pi}^{\pi} e^{ik\lambda}\, dZ(\lambda) = \int_{-\pi}^{\pi} e^{ik\lambda}\, dZ'(\lambda) + \Delta Z(0) \tag{44}$$

where

$$Z'(\lambda) = \begin{cases} Z(\lambda) & \text{if } \lambda < 0 \\ Z(\lambda) - \Delta Z(0) & \text{if } \lambda \geq 0. \end{cases} \tag{45}$$

Notice that

$$E\, dZ'(\lambda)\, \overline{dZ'(\mu)} = \delta_{\lambda,\mu}\, dF'(\lambda) \tag{46}$$

where

$$F'(\lambda) = \begin{cases} F(\lambda) & \text{if } \lambda < 0 \\ F(\lambda) - \Delta F(0) & \text{if } \lambda \geq 0 \end{cases} \tag{47}$$

is continuous at $\lambda = 0$ since the possible jump $\Delta F(0)$ of $F$ at zero has been removed. The expectation

$$E\left| \frac{1}{n} \sum_{k=1}^{n} X_k - m - \Delta Z(0) \right|^2 = E\left| \frac{1}{n} \sum_{k=1}^{n} \int_{-\pi}^{\pi} e^{ik\lambda} \, dZ'(\lambda) \right|^2$$

$$= \frac{1}{n^2} \int_{-\pi}^{\pi} \frac{\sin^2 \frac{n}{2} \lambda}{\sin^2 \frac{\lambda}{2}} \, dF'(\lambda) \tag{48}$$

$$\leq F'(\varepsilon) - F'(-\varepsilon) + \frac{1}{n^2 \sin^2 \frac{\varepsilon}{2}} [F'(\pi) - F'(-\pi)].$$

Since $F'$ is continuous at zero this last expression can be made arbitrarily small with $n$ sufficiently large. Therefore

$$\lim_{n \to \infty} \frac{1}{n} \sum_{k=1}^{n} X_k = \Delta Z(0) + m \tag{49}$$

in mean square. Notice that this indicates that the limiting time average $\lim_{n \to \infty} \frac{1}{n} \sum_{k=1}^{n} X_k$ is equal to the space average $m = EX_k$ if and only if $\Delta Z(0) = 0$.

## c. The Linear Prediction Problem and Autoregressive Schemes

Let us first consider the problem of predicting $X_0$ by a linear form in the observed variables $X_{-1}, \ldots, X_{-n}$. The object is to obtain a best linear predictor $X_0^* = \sum_{j=1}^{n} a_j X_{-j}$ in the sense that it attains the minimal mean square error of prediction among all linear predictors in terms of $X_{-1}, \ldots, X_{-n}$

$$E|X_0 - X_0^*|^2 = \min_{a_j} E\left|X_0 - \sum_{j=1}^{n} a_j X_{-j}\right|^2. \tag{1}$$

It is natural to call this a one-step prediction problem since we want to predict one step into the future. Predicting $X_1$ in terms of $X_{-1}$,

. . . , $X_n$ would be a two-step prediction problem. For convenience assume that $X_{-n}, X_{-n+1}, \ldots, X_0$ are a finite segment of a weakly stationary process $\{X_k\}$ with mean zero and a spectral distribution function $F(\lambda)$ with an infinite number of points of increase. $F$ is said to have a point of increase at $\mu$ if for every $\varepsilon > 0$, $F(\mu + \varepsilon) - F(\mu - \varepsilon) > 0$. The best linear predictor will be given in terms of an appropriately constructed set of orthogonal random variables. The procedure by which the orthogonal random variables are to be constructed is usually referred to as the Gramm-Schmidt orthogonalization procedure (see Apostol [1]). Let

$$\xi_{-n} = X_{-n}/\{E|X_{-n}|^2\}^{\frac{1}{2}}. \tag{2}$$

Then $E|\xi_{-n}|^2 = 1$. Set

$$\eta_{-n+1} = X_{-n+1} - E(X_{-n+1}\bar{\xi}_{-n})\xi_{-n} \tag{3}$$

and

$$\xi_{-n+1} = \eta_{-n+1}/\{E|\eta_{-n+1}|^2\}^{\frac{1}{2}}. \tag{4}$$

We shall later see that $\xi_{-n+1}$ is well defined since we shall show that $E|\eta_{-n+1}|^2 > 0$. Assuming this, notice that

$$E\xi_{-n}\bar{\xi}_{-n+1} = 0, \quad E|\xi_{-n+1}|^2 = 1. \tag{5}$$

Suppose that $\xi_{-n+i}$ has been constructed as a linear form in $X_{-n}, \ldots, X_{-n+i}$ in such a way that

$$E\xi_{-n+i}\bar{\xi}_{-n+i'} = \delta_{i,i'} \tag{6}$$

$i, i' = 0, 1, \ldots, j - 1$. The random variable $\xi_{-n+j}$ is constructed as follows. Let

$$\eta_{-n+j} = X_{-n+j} - \sum_{i=0}^{j-1} E(X_{-n+j}\bar{\xi}_{-n+i})\xi_{-n+i}. \tag{7}$$

Now $\eta_{-n+j}$ must have positive variance. For otherwise there would be a linear relationship among $X_{-n}, \ldots, X_{-n+j}$

$$\sum_{i=0}^{j} a_i X_{-n+i} = 0 \tag{8}$$

with the $a_i$'s not all zero so that

$$E\left| \sum_{i=0}^{j} a_i X_{-n+i} \right|^2 = \int_{-\pi}^{\pi} \left| \sum_{k=0}^{j} a_k e^{ik\lambda} \right|^2 dF(\lambda) = 0. \tag{9}$$

But this could not hold unless $F$ had at most a finite number of points of increase and we have assumed this is not the case. Thus $E|\eta_{-n+j}|^2 > 0$. Set

$$\xi_{-n+j} = \eta_{-n+j}/\{E|\eta_{-n+j}|^2\}^{\frac{1}{2}}. \tag{10}$$

In this way we recursively construct orthonormal random variables $\xi_{-n+i}$, $i = 0, 1, \ldots, n$,

$$E\xi_{-n+i}\bar{\xi}_{-n+j} = \delta_{i,j} \tag{11}$$

such that $\xi_{-n+i}$ is given by a linear form in $X_{-n}, \ldots, X_{-n+i}$. A possible predictor is a linear form in $X_{-n}, \ldots, X_{-1}$ and hence in terms of $\xi_{-n}, \ldots, \xi_{-1}$

$$X_0^* = \sum_{j=0}^{n-1} a_j X_{-n+j} = \sum_{j=0}^{n-1} \alpha_j \xi_{-n+j}. \tag{12}$$

The mean square error of prediction with this predictor is given by

$$E|X_0 - X_0^*|^2 = E|X_0 - \sum_{j=0}^{n-1} \alpha_j \xi_{-n+j}|^2$$

$$= E|X_0|^2 - 2 \operatorname{Re} \sum_{j=0}^{n-1} \bar{\alpha}_j E(X_0 \bar{\xi}_{-n+j}) + \sum_{j=0}^{n-1} |\alpha_j|^2 \tag{13}$$

$$= E|X_0|^2 - \sum_{j=0}^{n-1} |E(X_0 \bar{\xi}_{-n+j})|^2 + \sum_{j=0}^{n-1} |\alpha_j - E(X_0 \bar{\xi}_{-n+j})|^2.$$

It is clear that the mean square error of prediction is minimized by setting $\alpha_j = E(X_0 \bar{\xi}_{-n+j})$ so that the best linear predictor is given by

$$X_0^* = \sum_{j=0}^{n-1} E(X_0 \bar{\xi}_{-n+j}) \xi_{-n+j}. \tag{14}$$

The *error of prediction* $X_0 - X_0^*$ *is orthogonal to* $X_{-n}, \ldots, X_{-1}$ since $X_0 = \sum_{j=0}^{n} E(X_0 \bar{\xi}_{-n+j}) \xi_{-n+j}$. In fact, a little reflection indicates that the *best linear predictor is characterized by this orthogonality property.* For suppose $X_0 - X_0^*$ is orthogonal to $X_{-n}, \ldots, X_{-1}$ for some predictor $X_0^*$. Because of this orthogonality property it follows that

$$X_0 - X_0^* = \beta \xi_0 \tag{15}$$

for some constant $\beta$. We know that

$$X_0 = \sum_{j=0}^{n} E(X_0 \bar{\xi}_{-n+j}) \xi_{-n+j}. \tag{16}$$

Thus

$$X_0^* = \sum_{j=0}^{n} E(X_0 \bar{\xi}_{-n+j}) \xi_{-n+j} - \beta \xi_0. \tag{17}$$

Since $X_0^*$ is a predictor it must be a linear form in $X_{-n}, \ldots, X_{-1}$ or equivalently $\xi_{-n}, \ldots, \xi_{-1}$ and therefore $\beta = E(X_0 \bar{\xi}_0)$ so that $X_0^*$ is the best linear predictor (see (14)).

Let us now consider a weakly stationary process $X_n$ that satisfies the system of difference equations

$$\sum_{k=0}^{m} a_k X_{n-k} = \xi_n \tag{18}$$

where the $\xi_n$ are a sequence of orthogonal random variables

$$E\xi_n \bar{\xi}_m = \delta_{n,m}\sigma^2, \ \sigma^2 > 0, \tag{19}$$

with mean zero, $E\xi_n \equiv 0$, and $a_0 \neq 0$. The processes $\{X_n\}$ and $\{\xi_n\}$ are both stationary and hence have random spectral representations

$$X_n = \int_{-\pi}^{\pi} e^{in\lambda} \, dZ_x(\lambda)$$
$$\xi_n = \int_{-\pi}^{\pi} e^{in\lambda} \, dZ_\xi(\lambda). \tag{20}$$

Clearly

$$E \, dZ_\xi(\lambda) \, \overline{dZ_\xi(\mu)} = \delta_{\lambda,\mu} \frac{d\lambda}{2\pi} \sigma^2 \tag{21}$$

because of the orthogonality of the $\{\xi_n\}$ process. The difference equations (18) imply that

$$\int_{-\pi}^{\pi} e^{in\lambda} a(e^{-i\lambda}) \, dZ_x(\lambda) = \int_{-\pi}^{\pi} e^{in\lambda} \, dZ_\xi(\lambda) \tag{22}$$

where $a(z) = \sum_{k=0}^{m} a_k z^k$. On approximating the function which is 1 for $-\pi \leq \lambda \leq \mu$ and zero for $\mu < \lambda < \pi$ by linear forms $g_n(\lambda)$ in the complete system of functions $e^{in\lambda}$, $n = 0, \pm 1, \ldots$, and using property 2 of random integrals cited in section b, we obtain

$$\int_{-\pi}^{\mu} a(e^{-i\lambda}) \, dZ_x(\lambda) = Z_\xi(\mu). \tag{23}$$

From this, it follows that

$$\int_{-\pi}^{\mu} |a(e^{-i\lambda})|^2 \, dF_x(\lambda) = \frac{\sigma^2}{2\pi} (\mu + \pi) \tag{24}$$

where $F_x(\lambda)$ and $F_\xi(\lambda) = \frac{\sigma^2}{2\pi} (\lambda + \pi)$ are the spectral distribution functions of the $\{X_n\}$ and $\{\xi_n\}$ processes respectively. If $F_x(\lambda)$ has any jumps $a(e^{-i\lambda})$ must be zero there since the right-hand side of (24) is continuous. Let $F_{x,d}(\lambda)$ be the jump part of $F(\lambda)$ if any exists, that is,

$$F_{x,d}(\lambda) = \sum_{\lambda_j \leq \lambda} \Delta F_x(\lambda_j) \tag{25}$$

where $\Delta F_x(\lambda_j)$ is the jump of $F_x$ at $\lambda_j$. Call the continuous part of $F_x(\lambda)$, $F'_x(\lambda)$

$$F'_x(\lambda) = F_x(\lambda) - F_{x,d}(\lambda). \tag{26}$$

Then

$$\int_{-\pi}^{\mu} |a(e^{-i\lambda})|^2 \, dF'_x(\lambda) = \frac{\sigma^2}{2\pi} (\mu + \pi). \tag{27}$$

Since the right-hand side of (27) is absolutely continuous (with respect to Lebesgue measure) the left-hand side must be. Thus $F'_x(\lambda)$ is differentiable with

$$f_x(\lambda) = \frac{dF'_x(\lambda)}{d\lambda} = \frac{\sigma^2}{2\pi} \Big/ |a(e^{-i\lambda})|^2. \tag{28}$$

Since the spectral density function $f_x(\lambda)$ is integrable, $a(e^{-i\lambda})$ can have no zeros and jumps of $F_x$ cannot exist. *Processes satisfying a system of difference equations of the form (18) are called autoregressive schemes.* We have just shown that *a weakly stationary solution of this system of equations exists if and only if $a(e^{-i\lambda})$ has no zeros.* Further, *if a solution exists, it is unique. The spectral distribution function $F_x(\lambda)$ of the scheme is absolutely continuous with a spectral density given by*

$$f_x(\lambda) = \frac{dF_x(\lambda)}{d\lambda} = \frac{\sigma^2}{2\pi} \Big/ |a(e^{-i\lambda})|^2. \tag{29}$$

Notice that we have implicitly also shown that

$$X_n = \int_{-\pi}^{\pi} e^{in\lambda} \frac{1}{a(e^{-i\lambda})} \, dZ_\xi(\lambda). \tag{30}$$

One of the main reasons for interest in autoregressive schemes is due to the fact that the linear prediction problem is simple when dealing with them. Suppose that the difference equations (18) have been set up so that $\xi_n$ is orthogonal to $X_{n-1}, X_{n-2}, \ldots$ for all $n$. From the discussion at the beginning of this section

$$X_n^* = \sum_{k=1}^{m} \frac{a_k}{a_0} X_{n-k} \tag{31}$$

is the best linear (one-step) predictor of $X_n$ in terms of $X_{n-1}, X_{n-2}, \ldots$. The prediction error is $\xi_n/a_0$ with mean square error of prediction

$$\sigma^2/|a_0|^2. \tag{32}$$

The best linear predictor (two-step) of $X_n$ in terms of $X_{n-2}$, $X_{n-3}$, . . .
is obtained by replacing $X_{n-1}$ by $X_{n-1}^*$ in (31).

Let us now determine the circumstances under which $\xi_n$ is orthogonal to $X_{n-1}$, $X_{n-2}$, . . . . We have already seen that

$$X_n = \int_{-\pi}^{\pi} e^{in\lambda} \frac{1}{a(e^{-i\lambda})} \, dZ_\xi(\lambda) \tag{33}$$

where $a(z) = \sum_{k=0}^{m} a_k z^k$. Let the zeros of $a(z)$ in the complex plane be $z_1, z_2, \ldots, z_m$ where the zeros are enumerated so that $z_1, \ldots, z_p$ are inside and $z_{p+1}, \ldots, z_m$ are outside the unit circle $|z| = 1$. To avoid a more elaborate notation we assume that all zeros are simple. The following partial fraction expansion of $1/a(z)$

$$\frac{1}{a(z)} = \sum_{\nu=1}^{m} \frac{A_\nu}{z - z_\nu} = \sum_{\nu=1}^{p} \frac{A_\nu}{z} \sum_{j=0}^{\infty} \left(\frac{z_\nu}{z}\right)^j - \sum_{\nu=p+1}^{m} \frac{A_\nu}{z_\nu} \sum_{j=0}^{\infty} \left(\frac{z}{z_\nu}\right)^j \tag{34}$$

is uniformly convergent for $|z| = 1$. The argument is completely analogous for multiple roots except for a more complicated partial fraction expansion. Introducing (34) into (33) the following representation of $X_n$ as a moving average of $\xi_n$'s is obtained

$$X_n = \sum_{\nu=1}^{p} A_\nu \sum_{j=0}^{\infty} \xi_{n+j+1} z_\nu^j - \sum_{\nu=p+1}^{m} A_\nu \sum_{j=0}^{\infty} \xi_{n-j} z_\nu^{-j-1}. \tag{35}$$

It is easily seen that $\xi_n$ *will be orthogonal to* $X_{n-1}$, $X_{n-2}$, . . . *if and only if all the zeros* $z_\nu$ *are outside the unit circle, that* $|z_\nu| > 1$ *for all* $\nu$. This is a bit disconcerting because $a(z)$ will generally have zeros inside and outside the unit circle. However, we shall show that if $\{X_n\}$ is an autoregressive scheme of order $m$ (satisfies a system of difference equations of the form (18)), one can find a system of difference equations

$$\sum_{k=0}^{m} b_k X_{n-k} = \eta_n, \qquad b_0 \neq 0, \tag{36}$$

with

$$E\eta_n \equiv 0, \qquad E\eta_n \bar{\eta}_m = \delta_{n,m} \sigma^2, \qquad \sigma^2 > 0, \tag{37}$$

that $\{X_n\}$ satisfies where $\eta_n$ is orthogonal to $X_{n-1}$, $X_{n-2}$, . . . for all $n$. Since $\{X_n\}$ satisfies (18), it has the representation (33). Suppose that $a(z)$ has $p$ zeros $z_1, \ldots, z_p$ inside the unit circle. Let

$$b(z) = \sum_{0}^{m} b_k z^k = a(z) \prod_{\nu=1}^{p} \left(\frac{z\bar{z}_\nu - 1}{z - z_\nu}\right). \tag{38}$$

Notice that $b(z)$ has all its zeros outside the unit circle. The process $\{X_n\}$ has the representation

$$X_n = \int_{-\pi}^{\pi} e^{in\lambda} \frac{1}{b(e^{-i\lambda})} \, dZ_\eta(\lambda) \tag{39}$$

where

$$Z_\eta(\lambda) = \int_{-\pi}^{\lambda} \prod_{\nu=1}^{p} \left( \frac{e^{-i\lambda} - z_\nu}{e^{-i\lambda}\bar{z}_\nu - 1} \right) dZ_\xi(\lambda). \tag{40}$$

$Z_\eta(\lambda)$ is a process with orthogonal increments such that

$$E \, dZ_\eta(\lambda) \, \overline{dZ_\eta(\mu)} = \delta_{\lambda,\mu} \frac{\sigma^2}{2\pi} \tag{41}$$

since

$$\left| \frac{e^{-i\lambda} - z_\nu}{e^{-i\lambda}\bar{z}_\nu - 1} \right| = 1. \tag{42}$$

But this implies that

$$\sum_{k=0}^{m} b_k X_{n-k} = \int_{-\pi}^{\pi} e^{in\lambda} b(e^{-i\lambda}) \, dZ_x(\lambda)$$

$$= \int_{-\pi}^{\pi} e^{in\lambda} \, dZ_\eta(\lambda) = \eta_n \tag{43}$$

with $\eta_n$ orthogonal to $X_{n-1}, X_{n-2}, \ldots$ .

One can easily express the one-step prediction error $\sigma^2/|b_0|^2$ in terms of the spectral density of the autoregressive scheme. First notice that

$$\int_{-\pi}^{\pi} \log |e^{-i\lambda} - z|^2 \, d\lambda - 2\pi \log |z|^2$$

$$= \int_{-\pi}^{\pi} \log (1 - e^{-i\lambda}z^{-1})(1 - e^{i\lambda}\bar{z}^{-1}) \, d\lambda$$

$$= \int_{-\pi}^{\pi} \left\{ \sum_{k=1}^{\infty} e^{-ik\lambda} z^{-k}/k + \sum_{k=1}^{\infty} e^{ik\lambda} \bar{z}^{-k}/k \right\} d\lambda = 0 \tag{44}$$

if $|z| > 1$. But then

$$\frac{1}{2\pi} \int_{-\pi}^{\pi} \log f_x(\lambda) \, d\lambda - \log \left( \frac{\sigma^2}{2\pi} \right) = \frac{1}{2\pi} \int_{-\pi}^{\pi} \log |b(e^{-i\lambda})|^{-2} \, d\lambda$$

$$= -\frac{1}{2\pi} \sum_{\nu=1}^{m} \int_{-\pi}^{\pi} \log |e^{-i\lambda} - z_\nu|^2 \, d\lambda - \log |b_m|^2$$

$$= -\sum_{\nu=1}^{m} \log |z_\nu|^2 - \log |b_m|^2 = -\log |b_0|^2 \tag{45}$$

where the $z_\nu$'s are the zeros of $b$ and hence all have absolute value greater than one. *The one-step prediction error is thus given by*

$$\frac{\sigma^2}{|b_0|^2} = 2\pi \exp\left\{\frac{1}{2\pi} \int_{-\pi}^{\pi} \log f_x(\lambda)\, d\lambda\right\}. \tag{46}$$

This result is valid for all weakly stationary processes with absolutely continuous spectral distribution function (see [27]). We shall, however, prove this result only for such processes with spectral density positive and continuous on $[-\pi,\pi]$ and $f(-\pi) = f(\pi)$.

The reciprocal of the spectral density of an autoregressive scheme is a positive continuous trigonometric polynomial

$$\frac{\sigma^2}{2\pi}\frac{1}{f_x(\lambda)} = |b(e^{-i\lambda})|^2 = \sum_{k=0}^{m} A_k \cos k\lambda + \sum_{k=1}^{m} B_k \sin k\lambda. \tag{47}$$

We should like to show that *every positive continuous trigonometric polynomial has a representation as the absolute square of a one-sided trigonometric polynomial* $|\sum_{k=0}^{m} b_k e^{-ik\lambda}|^2$. This is a special case of a result due to Fejér and Riesz (see [27]). Let $p(\lambda)$ be such a trigonometric polynomial

$$p(\lambda) = \sum_{k=0}^{m} A_k \cos k\lambda + \sum_{k=1}^{m} B_k \sin k\lambda \tag{48}$$

with $A_0$ and either $A_m$ or $B_m$ not zero. Replace $\cos k\lambda$ by $(z^k + z^{-k})/2$ and $\sin k\lambda$ by $(z^k - z^{-k})/2i$. The resulting expression has the form $z^{-m}G(z)$ where $G(z)$ is a polynomial in $z$ of degree $2m$ with $G(0) \neq 0$. Further $\bar{G}\left(\frac{1}{z}\right) = G(z)$. But this means that there is a one-one correspondence between the roots of $G$ inside the unit circle and those outside. If $z_0$ is any root of $G(z)$ with $|z_0| < 1$, then $\bar{z}_0^{-1}$ is a root of $G$ outside the unit circle. Zero cannot be a root since $G(0) \neq 0$. Further there can be no roots of absolute value one since the polynomial $p(\lambda)$ is assumed to be positive. Thus $G$ has the form

$$G(z) = A \prod_{\nu=1}^{m} (z - z_\nu)(z - \bar{z}_\nu^{-1}) \tag{49}$$

and hence

$$p(\lambda) = Ae^{-im\lambda} \prod_{\nu=1}^{m} (e^{i\lambda} - z_\nu)(e^{i\lambda} - \bar{z}_\nu^{-1})$$

$$= B \prod_{\nu=1}^{m} |e^{i\lambda} - z_\nu|^2 \tag{50}$$

where

$$B = A \prod_{\nu=1}^{m} (-\bar{z}_\nu^{-1}) > 0. \tag{51}$$

The desired representation is given above in (50). Notice that this implies that any stationary process with spectral density the reciprocal of a trigonometric polynomial is an autoregressive scheme.

Now consider a stationary process $\{X_n\}$ with absolutely continuous spectral distribution function and spectral density $f(\lambda)$ continuous and positive on $[-\pi,\pi]$ with $f(-\pi) = f(\pi)$. The mean square error of prediction in predicting $X_n$ by a linear form in $X_{n-1}, \ldots, X_{n-m}$ is given by

$$\sigma_m^2(f(\lambda)) = \min_{c_j} E \left| X_n - \sum_{j=1}^{m} c_j X_{n-j} \right|^2$$

$$= \min_{c_j} \int_{-\pi}^{\pi} \left| 1 - \sum_{j=1}^{m} c_j e^{-ij\lambda} \right|^2 f(\lambda) \, d\lambda. \tag{52}$$

It is natural to call the limit of $\sigma_m^2(f(\lambda))$ as $m \to \infty$, the mean square error of prediction of $X_n$ when predicting by linear forms in terms of the past, that is, $X_{n-1}, X_{n-2}, \ldots$. We shall show that

$$\sigma^2(f(\lambda)) = \lim_{m \to \infty} \sigma_m^2(f(\lambda))$$

$$= 2\pi \exp \left\{ \frac{1}{2\pi} \int_{-\pi}^{\pi} \log f(\lambda) \, d\lambda \right\}. \tag{53}$$

Given any $\varepsilon > 0$, by the Weierstrass approximation theorem (see Problem 10 of Chapter II) one can find positive trigonometric polynomials $p_1(\cdot), p_2(\cdot)$ such that

$$f(\lambda) - \varepsilon < p_1(\lambda) \le f(\lambda) \le p_2(\lambda) < f(\lambda) + \varepsilon \tag{54}$$

for all $\lambda$. But it is clear from (52) that

$$\sigma^2(f(\lambda)) \le \sigma^2(g(\lambda)) \tag{55}$$

if $f(\lambda) \le g(\lambda)$ for all $\lambda$. Further, a stationary process with spectral density $p_i(\lambda)$ has mean square error of prediction

$$2\pi \exp \left\{ \frac{1}{2\pi} \int_{-\pi}^{\pi} \log p_i(\lambda) \, d\lambda \right\} = \sigma^2(p_i(\lambda)), \ i = 1, 2, \tag{56}$$

since it is an autoregressive scheme. This implies that

$$\sigma^2(p_1(\lambda)) \le \sigma^2(f(\lambda)) \le \sigma^2(p_2(\lambda)) \tag{57}$$

with $\sigma^2(p_i(\lambda))$, $i = 1, 2$, given by (56). This cannot be valid for every $\varepsilon > 0$ unless (53) holds.

# d. Spectral Estimates for Normal Processes

Let $\{X_n\}$ be a discrete parameter normal stationary process. If the mean $m \equiv EX_n$ of the process is not zero, we have already seen that the time average

$$m^* = \frac{1}{N} \sum_{j=1}^{N} X_j \tag{1}$$

will be a reasonable estimate of $m$ if there is no jump $\Delta Z(0)$ in the random spectral resolution of $X_n$ at $\lambda = 0$. For then $m^*$ will converge to $m$ in mean square as $N \to \infty$ by the mean square ergodic theorem (see section b of this chapter). A detailed discussion of how such an estimate compares with other linear estimates can be found in [25] and [26].

However, in many cases one is also interested in obtaining information about the spectrum of the process from a sample $X_n$, $n = 1$, . . . , $N$. Assume that the spectral distribution function of $X_n$ is absolutely continuous with a positive continuous spectral density on $[-\pi,\pi]$. In this section whenever we speak of continuity on $[-\pi,\pi]$, it is to be understood that $-\pi$ is identified with $\pi$ and hence that $f(\pi) = f(-\pi)$. In fact we are really dealing with the points on a circle rather than a line segment. Keeping this in mind, if the points $s$ with $|s - \pi| < \varepsilon$ are referred to, it is understood that one means the points $s$ with $-\pi \leq s < -\pi + \varepsilon$ and $\pi - \varepsilon < s \leq \pi$. Our object is to investigate estimates of $f(\lambda)$. If the covariance sequence $r_n = Ex_m x_{m+n}$ is absolutely summable, that is, $\Sigma|r_k| < \infty$, the spectral density $f(\lambda)$ is given by

$$f(\lambda) = \frac{1}{2\pi} \sum_{n=-\infty}^{\infty} r_n e^{-in\lambda}. \tag{2}$$

One approach is to replace $r_n$ wherever possible by a good estimate $r_n(N)$. A good estimate of $r_n$ is given by

$$r_n(N) = \frac{1}{N} \sum_{m=1}^{N-|n|} X_m X_{m+|n|} \tag{3}$$

if $|n| < N$. Obviously, if $|n| \geq N$, $r_n$ cannot be estimated from the sample since there are no lags $n$ that large available in the sample. One might as well estimate them by zero. The resulting estimate is

$$
\begin{aligned}
I_N(\lambda) &= \frac{1}{2\pi} \sum_{n=-N}^{N} r_n(N) e^{-in\lambda} \\
&= \frac{1}{2\pi N} \left| \sum_{j=1}^{N} X_j e^{-ij\lambda} \right|^2,
\end{aligned}
\tag{4}
$$

commonly called the *periodogram*.

First consider the bias of $I_N(\lambda)$ as an estimate of $f(\lambda)$, that is,

$$
b_N(\lambda) = EI_N(\lambda) - f(\lambda).
\tag{5}
$$

The bias indicates how well the estimate is centered. Now

$$
\begin{aligned}
EI_N(\lambda) &= E \frac{1}{2\pi N} \left| \sum_{j=1}^{N} X_j e^{-ij\lambda} \right|^2 \\
&= \frac{1}{2\pi N} E \left| \int_{-\pi}^{\pi} \sum_{1}^{N} e^{ij(\mu-\lambda)} \, dZ(\mu) \right|^2 \\
&= \frac{1}{2\pi N} \int_{-\pi}^{\pi} \frac{\sin^2 \frac{N}{2}(\mu-\lambda)}{\sin^2 \frac{\mu-\lambda}{2}} f(\mu) \, d\mu.
\end{aligned}
\tag{6}
$$

The bias is therefore

$$
b_N(\lambda) = \frac{1}{2\pi N} \int_{-\pi}^{\pi} \frac{\sin^2 \frac{N}{2}(\mu-\lambda)}{\sin^2 \frac{\mu-\lambda}{2}} [f(\mu) - f(\lambda)] \, d\mu.
\tag{7}
$$

Notice that the weight function

$$
\frac{1}{2\pi N} \frac{\sin^2 \frac{N}{2}\lambda}{\sin^2 \frac{\lambda}{2}}
\tag{8}
$$

is non-negative and it has integral one over $[-\pi,\pi]$. Further, given any $\varepsilon > 0$

$$\lim_{N\to\infty} \int_{|\lambda|\leq\varepsilon} \frac{1}{2\pi N} \frac{\sin^2 \frac{N}{2}\lambda}{\sin^2 \frac{\lambda}{2}} \, d\lambda = 1. \tag{9}$$

Thus, all the mass of the weight function accumulates in the immediate neighborhood of $\lambda = 0$ as $N \to \infty$. This implies that $I_N(\lambda)$ is asymptotically unbiased since $b_N(\lambda) \to 0$ as $N \to \infty$. If the spectral density is assumed to be sufficiently smooth, one can get convenient bounds on the rate at which $b_N(\lambda) \to 0$. For example, suppose that $f(\lambda)$ is continuously differentiable. Now $f(\mu) - f(\lambda)$ is given by

$$f(\mu) - f(\lambda) = f'(\lambda + \theta(\mu - \lambda))(\mu - \lambda) \tag{10}$$

$|\theta| < 1$, for $|\mu - \lambda| < A < \frac{\pi}{2}$. The following bound is simply obtained by using formula (10)

$$|b_N(\lambda)| \leq \frac{\max |f'|}{2\pi N} \left\{ \int_{|\mu-\lambda|<\frac{A}{N}} + \int_{\frac{A}{N}\leq|\mu-\lambda|<A} \right\} \frac{\sin^2 \frac{N}{2}(\mu - \lambda)}{\sin^2 \frac{1}{2}(\mu - \lambda)} |\mu - \lambda| \, d\mu$$

$$+ \frac{1}{2\pi N} \int_{|\mu-\lambda|\geq A} \frac{\sin^2 \frac{N}{2}(\mu - \lambda)}{\sin^2 \frac{1}{2}(\mu - \lambda)} |f(\mu) - f(\lambda)| \, d\mu \leq K \frac{\log N}{N} \tag{11}$$

where $K$ is a constant. Thus $b_N(\lambda)$ approaches zero with order of magnitude $\frac{\log N}{N}$ as $N \to \infty$. The bias of the periodogram is quite small. But, as we shall see, the variance of the periodogram is unfortunately large. Nonetheless a detailed estimate of the variance will be made since it will be useful in considering more reasonable estimates of the spectral density. We assume the spectral density $f(\lambda)$ to be continuously differentiable as we did in the estimate of the bias carried out above. The procedure we use is in part a modification of an argument carried out by P. Scheinok in his thesis [72]. Now

$$4\pi^2 N^2 \text{ cov } (I_N(\lambda), I_N(\mu)) = \sum_{\nu_i=1}^{N} e^{i\nu_1\lambda} e^{-i\nu_2\lambda} e^{i\nu_3\mu} e^{-i\nu_4\mu} \text{ cov } (x_{\nu_1} x_{\nu_2}, x_{\nu_3} x_{\nu_4}) \tag{12}$$

where

$$\text{cov } (x_{\nu_1} x_{\nu_2}, x_{\nu_3} x_{\nu_4}) = r_{\nu_1-\nu_3} r_{\nu_2-\nu_4} + r_{\nu_1-\nu_4} r_{\nu_2-\nu_3} \tag{13}$$

(see Problem 10 of Chapter IV). Using the spectral representation of the covariances, we note that

$$4\pi^2 N^2 \operatorname{cov}\left(I_N(\lambda), I_N(\mu)\right)$$

$$= \int\int_{-\pi}^{\pi} \sum_{\nu_i=1}^{N} \{e^{i\nu_1(\lambda+\alpha)}e^{i\nu_3(\mu-\alpha)}e^{-i\nu_2(\lambda-\beta)}e^{-i\nu_4(\mu+\beta)}$$

$$+ e^{i\nu_1(\lambda+\alpha)}e^{-i\nu_4(\mu+\alpha)}e^{-i\nu_2(\lambda-\beta)}e^{i\nu_3(\mu-\beta)}\}f(\alpha)f(\beta)\,d\alpha\,d\beta \quad (14)$$

$$= \int\int_{-\pi}^{\pi} D_N(\lambda+\alpha)D_N(\mu-\alpha)D_N(-\lambda+\beta)D_N(-\mu-\beta)f(\alpha)f(\beta)\,d\alpha\,d\beta$$

$$+ \int\int_{-\pi}^{\pi} D_N(\lambda+\alpha)D_N(-\mu-\alpha)D_N(-\lambda+\beta)D_N(\mu-\beta)f(\alpha)f(\beta)\,d\alpha\,d\beta$$

$$\tag{15}$$

where

$$D_N(\lambda) = \sum_{\nu=1}^{N} e^{i\nu\lambda} = e^{i\lambda}\frac{e^{iN\lambda}-1}{e^{i\lambda}-1}. \tag{16}$$

If $f$ is set equal to the constant function one, the first integral and second integral of (15) are seen to be

$$(2\pi)^2\,\frac{\sin^2\frac{N}{2}(\lambda+\mu)}{\sin^2\frac{(\lambda+\mu)}{2}} \tag{17}$$

and

$$(2\pi)^2\,\frac{\sin^2\frac{N}{2}(\lambda-\mu)}{\sin^2\frac{(\lambda-\mu)}{2}} \tag{18}$$

respectively. This suggests estimating (15) by

$$(2\pi)^2\left\{\frac{\sin^2\frac{N}{2}(\lambda+\mu)}{\sin^2\frac{(\lambda+\mu)}{2}}+\frac{\sin^2\frac{N}{2}(\lambda-\mu)}{\sin^2\frac{(\lambda-\mu)}{2}}\right\}f(\lambda)f(\mu). \tag{19}$$

The difference between (15) and (19) is given by

$$
\iint\limits_{-\pi}^{\pi} D_N(\lambda + \alpha)D_N(\mu - \alpha)D_N(-\lambda + \beta)D_N(-\mu - \beta)[f(\alpha)f(\beta)
$$
$$
- f(\lambda)f(\mu)]\, d\alpha\, d\beta \quad (20)
$$
$$
+ \iint\limits_{-\pi}^{\pi} D_N(\lambda + \alpha)D_N(-\mu - \alpha)D_N(-\lambda + \beta)D_N(\mu - \beta)[f(\alpha)f(\beta)
$$
$$
- f(\lambda)f(\mu)]\, d\alpha\, d\beta.
$$

We shall only carry out the estimation for the second term of (20) since the estimation for the first term is completely analogous and leads to the same bound. By the Schwarz inequality

$$
\frac{1}{4\pi^2 N^2}\left|\iint\limits_{-\pi}^{\pi} D_N(\lambda + \alpha)D_N(-\mu - \alpha)D_N(-\lambda + \beta)D_N(\mu - \beta)[f(\alpha)f(\beta)\right.
$$
$$
\left. - f(\lambda)f(\mu)]\, d\alpha\, d\beta\right|
$$
$$
\quad (21)
$$
$$
\leq \frac{1}{4\pi^2 N^2}\left[\iint\limits_{-\pi}^{\pi} |D_N(\lambda + \alpha)D_N(\mu - \beta)|^2\ f(\alpha)f(\beta) - f(\lambda)f(\mu)|\, d\alpha\, d\beta\right.
$$
$$
\left.\iint\limits_{-\pi}^{\pi} |D_N(-\mu - \alpha)D_N(-\lambda + \beta)|^2\ f(\alpha)f(\beta) - f(\lambda)f(\mu)|\, d\alpha\, d\beta\right]^{\frac{1}{2}}.
$$

It will be enough to look at

$$
\frac{1}{4\pi^2 N^2}\iint\limits_{-\pi}^{\pi} D_N^2(\lambda + \alpha)D_N^2(\mu - \beta)|f(\alpha)f(\beta) - f(\lambda)f(\mu)|\, d\alpha\, d\beta. \quad (22)
$$

Divide the integral over $-\pi \leq \alpha, \beta \leq \pi$ into integrals over the two subdomains $R_1, R_2$ where

$$
\begin{aligned}
R_1 &= \{\alpha,\beta|\ |\lambda + \alpha| \geq A \text{ or } |\mu - \beta| \geq A\}\\
R_2 &= \{\alpha,\beta|\ |\lambda + \alpha| < A \text{ and } |\mu - \beta| < A\}.
\end{aligned} \quad (23)
$$

The integral over $R_1$ is bounded by $K_1/N$ where $K_1$ is some constant. By the mean value theorem

$$
f(\alpha)f(\beta) - f(\lambda)f(\mu) = f'(\bar{\alpha})f(\bar{\beta})(\alpha + \lambda) + f'(\bar{\beta})f(\bar{\alpha})(\beta - \mu) \quad (24)
$$

where $(\bar\alpha, \bar\beta)$ is a point on the line joining $(\alpha, \beta)$ to $(-\lambda, \mu)$. The integral over $R_2$ is seen to be bounded by $K_2 \log N / N$ by using this expansion and an estimation procedure like that leading to (11). Thus *the covariance of $I_N(\lambda)$ and $I_N(\mu)$ is given by*

$$\operatorname{cov}\,(I_N(\lambda), I_N(\mu)) = \left\{ \frac{\sin^2 \dfrac{N}{2}(\lambda - \mu)}{\sin^2 \dfrac{(\lambda - \mu)}{2}} + \frac{\sin^2 \dfrac{N}{2}(\lambda + \mu)}{\sin^2 \dfrac{(\lambda + \mu)}{2}} \right\} \frac{f(\lambda)f(\mu)}{N^2}$$
$$+ O\left(\frac{\log N}{N}\right) \tag{25}$$

*under the assumption of continuous differentiability of the spectral density.* Notice that the limiting variance

$$\lim_{N \to \infty} \sigma^2(I_N(\lambda)) = \begin{cases} f^2(\lambda) & \text{if } \lambda \neq 0, \pm\pi \\ 2f^2(\lambda) & \text{otherwise.} \end{cases} \tag{26}$$

This implies that the periodogram $I_N(\lambda)$ doesn't even converge to $f(\lambda)$ in mean square as $N \to \infty$ if $f(\lambda) \neq 0$, a dubious feature for a proposed estimate of the spectral density. However, $I_N(\lambda)$ and $I_N(\mu)$ are asymptotically uncorrelated if $\lambda \neq \mu$ and $f(\lambda)$ and $f(\mu)$ are positive. This suggests that we might get a reasonable estimate of $f(\lambda)$ by averaging $I_N(\mu)$ in the neighborhood of $\lambda$. Let $w_N(\mu)$ be a sequence of positive weight functions on $[-\pi, \pi]$ with the following properties:

(a) $\displaystyle \int_{-\pi}^{\pi} w_N(\mu)\, d\mu = 1$                             (27)

(b) $w_N(\mu) \to 0$ *uniformly in $\mu$ for $|\mu| \geq \varepsilon$*      (28)

*as $N \to \infty$ for every fixed $\varepsilon > 0$*

$$\text{(c)}\ \max_{|\mu| \leq \frac{A}{N}} \left| \frac{\displaystyle\int_{-\pi}^{\pi} w_N(\lambda) w_N(\lambda + \mu)\, d\lambda}{\displaystyle\int_{-\pi}^{\pi} w_N^2(\lambda)\, d\lambda} - 1 \right| \to 0 \tag{29}$$

*as $N \to \infty$ for an $A > 0$.*

Notice that (a) and (b) imply that more and more of the mass of the weight functions accumulate in the immediate vicinity of $\lambda = 0$ as $N \to \infty$ and hence that $\int_{-\pi}^{\pi} w_N^2(\alpha)\, d\alpha \to \infty$ as $N \to \infty$. The proposed sequence of estimates is

$$f_N(\lambda) = \int_{-\pi}^{\pi} w_N(\lambda - \mu) I_N(\mu)\, d\mu. \tag{30}$$

Our object is to investigate the bias and variance of this sequence of estimates as $N \rightarrow \infty$. From (6) it is clear that the bias $b_N(\lambda)$ is given by

$$b_N(\lambda) = Ef_N(\lambda) - f(\lambda)$$

$$= \int_{-\pi}^{\pi} w_N(\lambda - \mu) \int_{-\pi}^{\pi} \frac{1}{2\pi N} \frac{\sin^2 \frac{N}{2}(\mu - \alpha)}{\sin^2 \frac{\mu - \alpha}{2}} [f(\alpha) - f(\lambda)] \, d\alpha \, d\mu$$

$$= \int_{-\pi}^{\pi} w_N(\lambda - \mu)[f(\mu) - f(\lambda)] \, d\mu + O\left(\frac{\log N}{N}\right) \qquad (31)$$

under the assumption of continuous differentiability of the spectral density. Conditions (a), (b) on the sequence of weight functions imply that $b_N(\lambda) \rightarrow 0$ as $N \rightarrow \infty$ so that the estimates are asymptotically unbiased.

Consider now the variance $\sigma^2(f_N(\lambda))$ of the estimate $f_N(\lambda)$. Formula (25) implies that the variance is given by

$$\sigma^2(f_N(\lambda)) = \iint_{-\pi}^{\pi} w_N(\lambda - \alpha)w_N(\lambda - \beta) \, \text{cov} \, (I_N(\alpha), I_N(\beta)) \, d\alpha \, d\beta$$

$$= \frac{1}{N^2} \iint_{-\pi}^{\pi} w_N(\lambda - \alpha)w_N(\lambda - \beta) \frac{\sin^2 \frac{N}{2}(\alpha - \beta)}{\sin^2 \frac{(\alpha - \beta)}{2}} f(\alpha)f(\beta) \, d\alpha \, d\beta$$

$$+ \frac{1}{N^2} \iint_{-\pi}^{\pi} w_N(\lambda - \alpha)w_N(\lambda - \beta) \frac{\sin^2 \frac{N}{2}(\alpha + \beta)}{\sin^2 \frac{(\alpha + \beta)}{2}} f(\alpha)f(\beta) \, d\alpha \, d\beta$$

$$+ O\left(\frac{\log N}{N}\right). \qquad (32)$$

Assumption (c) essentially states that $w_N(\mu)$ accumulates mass in the neighborhood of $\mu = 0$ at a slower rate than the Fejér kernel. This suggests that one ought to approximate the two integrals of (32) by

$$\frac{2\pi}{N} \int_{-\pi}^{\pi} w_N^2(\lambda - \alpha)f^2(\alpha) \, d\alpha \qquad (33)$$

and

$$\frac{2\pi}{N} \int_{-\pi}^{\pi} w_N(\lambda - \alpha)w_N(\lambda + \alpha)f^2(\alpha) \, d\alpha \qquad (34)$$

respectively. It will only be necessary to carry out the approximation for the first integral in detail since the approximation is quite analogous for the second integral. Let us split

$$\frac{1}{N^2} \iint_{-\pi}^{\pi} w_N(\lambda - \alpha)w_N(\lambda - \beta) \frac{\sin^2 \dfrac{N}{2}(\alpha - \beta)}{\sin^2 \dfrac{(\alpha - \beta)}{2}} f(\alpha)f(\beta)\,d\alpha\,d\beta \quad (35)$$

into two integrals $I_1$ and $I_2$, the first over the region $|\alpha - \beta| \leq \dfrac{A}{N}$ and the second over $|\alpha - \beta| > \dfrac{A}{N}$. We have

$$I_1 = \frac{1}{N^2} \int_{|u| \leq \frac{A}{N}} \frac{\sin^2 \dfrac{N}{2} u}{\sin^2 \dfrac{u}{2}} \int_{-\pi}^{\pi} w_N(\lambda - \alpha)w_N(\lambda - \alpha + u)f(\alpha)f(\alpha - u)\,d\alpha\,du$$

$$= \frac{1}{N^2} \int_{|u| \leq \frac{A}{N}} \frac{\sin^2 \dfrac{Nu}{2}}{\sin^2 \dfrac{u}{2}} \int_{-\pi}^{\pi} w_N^2(\lambda - \alpha)f^2(\alpha)\,d\alpha\,du + o\left(\frac{1}{N}\int_{-\pi}^{\pi} w_N^2(\alpha)\,d\alpha\right)$$

$$= \frac{2\pi}{N} \int_{-\pi}^{\pi} w_N^2(\lambda - \alpha)f^2(\alpha)\,d\alpha + o\left(\frac{1}{N}\int_{-\pi}^{\pi} w_N^2(\alpha)\,d\alpha\right). \quad (36)$$

Similarly

$$I_2 = \frac{1}{N^2} \int_{|u| > \frac{A}{N}} \frac{\sin^2 \dfrac{Nu}{2}}{\sin^2 \dfrac{u}{2}} \int_{-\pi}^{\pi} w_N(\lambda - \alpha)w_N(\lambda - \alpha + u)f(\alpha)f(\alpha - u)\,d\alpha\,du$$

$$\leq \frac{K^2}{N^2} \int_{|u| > \frac{A}{N}} \frac{1}{\sin^2 \dfrac{u}{2}} \int_{-\pi}^{\pi} w_N^2(\alpha)\,d\alpha\,du$$

$$\leq \frac{K'}{NA} \int_{-\pi}^{\pi} w_N^2(\alpha)\,d\alpha = o\left(\frac{1}{N}\int_{-\pi}^{\pi} w_N^2(\alpha)\,d\alpha\right) \quad (37)$$

since $A$ can be made arbitrarily large. Thus

$$\sigma^2(f_N(\lambda)) \cong \frac{2\pi}{N} \int_{-\pi}^{\pi} w_N^2(\lambda - \alpha)f^2(\alpha)\,d\alpha$$

$$+ \frac{2\pi}{N} \int_{-\pi}^{\pi} w_N(\lambda - \alpha)w_N(\lambda + \alpha)f^2(\alpha)\,d\alpha \quad (38)$$

as $N \to \infty$. If $w_N(\lambda)$ is symmetric about $\lambda = 0$ as it is in most applications, a further approximation leads us to

$$\sigma^2(f_N(\lambda)) \cong \begin{cases} \dfrac{2\pi}{N} f^2(\lambda) \displaystyle\int_{-\pi}^{\pi} w_N^2(\alpha)\, d\alpha & \text{if } \lambda \neq 0, \pm\pi \\[2mm] \dfrac{4\pi}{N} f^2(\lambda) \displaystyle\int_{-\pi}^{\pi} w_N^2(\alpha)\, d\alpha & \text{otherwise.} \end{cases} \qquad (39)$$

A few simple weight functions are given by

$$w_N(\lambda) = \frac{1}{2\pi h_N} \frac{\sin^2 \dfrac{h_N}{2}\lambda}{\sin^2 \dfrac{\lambda}{2}} \qquad (40)$$

and

$$w_N(\lambda) = \begin{cases} \dfrac{1}{2h_N} & |\lambda| \leq h_N \\[2mm] 0 & \text{otherwise} \end{cases} \qquad (41)$$

where $h_N$ diverges at a rate slower than $N$ as $N \to \infty$.

Various approximations have been proposed for the probability distribution of such spectral estimates. The most commonly suggested is that with density function $g(x)$ where

$$g(x) = \begin{cases} \dfrac{\alpha^\lambda}{\Gamma(\lambda)} x^{\lambda-1} e^{-\alpha x} & \text{for } x > 0 \\[2mm] 0 & \text{otherwise.} \end{cases} \qquad (42)$$

Here $\alpha$, $\lambda$ are determined by fitting the first two moments. This approximation is quite reasonable for moderate sample size. A Gaussian distribution can be used when the sample size is large. A detailed discussion of approximations for small sample size can be found in [20] and [28].

The discussion of corresponding estimates in the continuous parameter case is similar. Extensive discussions of spectral analysis can be found in [6], [26], and [71].

In many cases the discrete parameter time series is obtained by discretely sampling a continuous parameter time series (see the discussion in section a of this chapter). The following conditions on the spectrum of the continuous parameter process are enough to ensure that the spectrum of the derived discrete parameter process satisfy the conditions assumed in this section. Let the spectrum of the continuous time parameter process be absolutely continuous with a continuously differentiable spectral density $f(\lambda)$. Further, let

$$f(\lambda) = O(|\lambda|^{-1-\varepsilon}) \tag{43}$$

as $|\lambda| \to \infty$ where $\varepsilon$ is some positive number. The derivative of the spectral density of the derived discrete parameter process is then continuous and is given by

$$\sum_{j=-\infty}^{\infty} f'\left(\lambda + \frac{2j\pi}{\Delta}\right) \tag{44}$$

where $\Delta$ is the sampling interval.

## e. Problems

1. Let $Y$ be a random variable with distribution function $F(y)$. Let $X_n = \exp\,(inY)$, $n = 0,\ \pm 1,\ \ldots\ .$ Compute the moments $EX_n\bar{X}_m$.

2. Let $X_n$, $n = 0,\ \pm 1,\ \pm 2,\ \ldots\ ,$ $EX_n = 0$, be a weakly stationary process so that $r_n = EX_kX_{k+n}$ has the representation $r_n = \int_{-\pi}^{\pi} e^{in\lambda}\,dF(\lambda)$ where $F$ is a bounded nondecreasing function. Show that $X_n$ has the representation $X_n = \int_{-\pi}^{\pi} e^{in\lambda}\,dZ(\lambda)$ where $Z(\lambda)$ is a process of orthogonal increments with $E\,dZ(\lambda)\,\overline{dZ(\mu)} = \delta_{\lambda,\mu}\,dF(\lambda)$.

3. Let $X_t$, $t\epsilon[0,2\pi]$, $EX_t \equiv 0$ be weakly stationary on the unit circle. This means that $t = 0$ is identified with $t = 2\pi$ and

$$EX_t\,\bar{X}_s = r_u \qquad u = t - s \bmod 2\pi.$$

Show that

$$r_t = \sum_{n=-\infty}^{\infty} a_n e^{itn}$$

where the numbers $a_n$ are the Fourier coefficients of $r_t$ and that

$$X_t = \sum_{n=-\infty}^{\infty} Z_n e^{itn}$$

with $EZ_n\bar{Z}_m = \delta_{n,m}a_n$.

**4.** A weakly stationary process $X_t$, $-\infty < t < \infty$, $EX_t \equiv 0$, has the representation

$$X_t = \int_{-\infty}^{\infty} e^{it\lambda} \, dZ(\lambda)$$

where $Z(\lambda)$ is a process with orthogonal increments. Show that a real-valued weakly stationary process has the representation

$$X_t = \int_0^{\infty} \cos t\lambda \, dZ_1(\lambda) + \int_0^{\infty} \sin t\lambda \, dZ_2(\lambda)$$

where

$$E \, dZ_j(\lambda) \, dZ_k(\mu) = \delta_{j,k} 2 \, dF(\lambda), \, j, \, k = 1, \, 2.$$

Further $dZ_1(\lambda) = 2 \, Re \, dZ(\lambda), \qquad dZ_2(\lambda) = -2 \, Im \, dZ(\lambda)$.

**5.** Let $X_t$, $-\infty < t < \infty$, $EX_t \equiv 0$, be a weakly stationary process. Assume that the spectrum of the process is band-limited to $[-\pi w, \pi w]$ so that $X_t$ has the representation

$$X_t = \int_{-\pi w}^{\pi w} e^{it\lambda} \, dZ(\lambda).$$

Show that

$$X_t = \sum_{n=-\infty}^{\infty} X_{n/w} \frac{\sin \pi(wt - n)}{\pi(wt - n)}.$$

**6.** Let $X_t$, $t = 0, \pm 1, \ldots$, $EX_t = 0$, be a strictly stationary Markov chain with a finite number of states and transition probability matrix $\mathbf{M}$. Assume that $\mathbf{M}$ has only simple eigenvalues. Find the spectral distribution function of $X_t$ in terms of the eigenvalues of $\mathbf{M}$.

**7.** Let $X_n$, $n = 0, \pm 1, \pm 2, \ldots$ be a normal process satisfying equations of the form

$$\sum_{k=0}^{m} a_k X_{n-k} = \sum_{k=0}^{m} b_k \xi_{n-k} \text{ with } a_0, \, b_0 \neq 0,$$

where the $\xi_n$ are independent normal random variables with mean zero and variance one. Under what conditions will there be a stationary solution $X_n$? Can you express the $X_n$ process directly in terms of the $\xi_n$ process? What is the spectrum of a stationary solution if it exists?

**8.** Let $\mathbf{X}_n = (X_n^{(1)}, X_n^{(2)})$, $EX_n \equiv 0$, $n = 0, \pm 1, \ldots$ be a weakly stationary two-vector-valued process, that is, $\mathbf{r}_{n,m} = E\mathbf{X}_n' \mathbf{X}_m$

depends only on the time difference $n - m$. Show that then $\mathbf{r}_{n-m} = \mathbf{r}_{n,m}$ has the representation $\mathbf{r}_{n-m} = \int_{-\pi}^{\pi} e^{i(n-m)\lambda} \, d\mathbf{F}(\lambda)$ where $\mathbf{F}(\lambda)$ is a $2 \times 2$ matrix-valued Hermitian function that is bounded (every element a bounded function) and nondecreasing in that $\Delta\mathbf{F}(\lambda) = \mathbf{F}(\lambda_1) - \mathbf{F}(\lambda_2)$ is a positive definite matrix for every pair $\lambda_1, \lambda_2$ with $\lambda_1 > \lambda_2$.

9.  Let $\mathbf{X}_n$ be a two-vector-valued weakly stationary process as in the last example. Assume that the components are real-valued. It then follows that $d\mathbf{F}(\lambda) = \overline{d\mathbf{F}(-\lambda)}$. Show that $\mathbf{X}_n$ has the real representation

$$X_n^{(1)} = \int_0^\pi \cos n\lambda \, dZ_1^{(1)}(\lambda) + \int_0^\pi \sin n\lambda \, dZ_2^{(1)}(\lambda)$$
$$X_n^{(2)} = \int_0^\pi \cos n\lambda \, dZ_1^{(2)}(\lambda) + \int_0^\pi \sin n\lambda \, dZ_2^{(2)}(\lambda)$$

where the $Z_i^{(j)}(\lambda)$ are real-valued processes with orthogonal increments such that

$$E \, dZ_i^{(1)}(\lambda) \, dZ_j^{(1)}(\mu) = E \, dZ_i^{(2)}(\lambda) \, dZ_j^{(2)}(\mu) = 2\delta_{i,j}\delta_{\lambda,\mu} \, dF_{i,j}(\lambda)$$
$$E \, dZ_i^{(1)}(\lambda) \, dZ_i^{(2)}(\mu) = 2\delta_{\lambda,\mu} \, Re \, dF_{1,2}(\lambda) \quad i, j = 1, 2$$
$$E \, dZ_1^{(1)}(\lambda) \, dZ_2^{(2)}(\mu) = -E \, dZ_2^{(1)}(\lambda) \, dZ_1^{(2)}(\mu) = 2\delta_{\lambda,\mu} \, Im \, dF_{1,2}(\lambda).$$

If the process has a spectral density function, that is, $\mathbf{F}(\lambda) = \int_{-\pi}^{\lambda} \mathbf{f}(\mu) \, d\mu$, the real and imaginary parts of the cross-spectral density $Re f_{12}(\lambda)$, $Im f_{12}(\lambda)$ are sometimes referred to as the cospectrum and quadrature spectrum of $X_n^{(1)}$ and $X_n^{(2)}$.

10.  Consider the results of the two previous examples in the case of a normal stationary two-vector-valued process $\mathbf{X}_n$ satisfying the matrix equation

$$\mathbf{X}_n = \mathbf{A}\mathbf{X}_{n-1} + \boldsymbol{\xi}_n$$

where $\boldsymbol{\xi}_n$ is a two-vector-valued process with components two independent normal white noise processes and $\mathbf{A}$ a $2 \times 2$ matrix with real elements such that $\mathbf{A}\mathbf{A}' \leq \mathbf{I}$.

11.  Consider what parallels of the results of this chapter you can obtain in the case of a two-parameter process $X_{t,\tau}$, $EX_{t,\tau} \equiv 0$, weakly stationary in that $EX_{t,\tau}X_{t+\alpha,\tau+\beta}$ depends only on the differences $\alpha, \beta$.

# Notes

1. For a more extensive discussion of weakly stationary processes one should refer to the books of Grenander and Rosenblatt [26] and Doob [12]. A general method of obtaining the harmonic representation of a stationary process and many allied results by Hilbert space methods is given in Karhunen's paper [38] on linear methods in probability theory. The book of Iu. Rosanov [A13] on stationary processes develops many of the basic results in this area.

2. The result on time averages $\frac{1}{n} \sum_{k=1}^{n} X_k$ of a weakly stationary process is essentially the von Neumann ergodic theorem (see [35]).

3. The linear prediction problem is considered in section c only for weakly stationary processes with a continuous spectral density with $f(-\pi) = f(\pi)$ bounded away from zero. A discussion of the general result and related material can be found in the excellent book of Grenander and Szegö [27] on Toeplitz forms. Recently there has been a good deal of work on the prediction problem for vector-valued weakly stationary time series (see [33], [56]). The initial work on prediction problems is due to Kolmogorov [46] and Wiener [77]. The analytic problem that basically enters into the prediction problem for discrete time parameter weakly stationary processes was treated by G. Szegö [75].

4. An extensive discussion of spectral analysis of stationary time series can be found in Grenander and Rosenblatt [26]. The derivation of the asymptotic properties of spectral estimates given in section d is special since it is carried out only for normal processes and under rather strong regularity restrictions on the spectral density of the process. However, it was felt that there are advantages to this since the derivation is elementary and is carried out completely in the spectral domain rather than in terms of the covariances of the process. Further, some of the intermediate estimates made such as (d. 32) give greater insight into the discontinuity in the limit formula at 0 and $\pm \pi$. For a derivation under more general conditions see Parzen's paper [59]. A discussion of spectral analysis for vector-valued stationary time series can be found in [24] and [71]. The books of T. W. Anderson [A1] and E. J. Hannan [A7] which have recently appeared are also very useful references.

# VIII

MARTINGALES

## a. Definition and Illustrations

We consider a family of processes called martingales. Though a number of other illustrative examples will be given, the most immediate interpretation is that of a fair game. The discrete parameter case is considered for simplicity. The parameter $n$ is assumed to run over all the integers, the positive integers or the negative integers. Let $X_n$ be a sequence of random variables with $\mathcal{B}_n$ a corresponding nondecreasing sequence of Borel subfields of $\mathcal{F}$. The random variable $X_n$ is assumed to be measurable with respect to $\mathcal{B}_n$. In fact, $\mathcal{B}_n$ is often taken to be the Borel field generated by $X_k$, $k \leq n$, though this is not necessarily the case in our discussion. Also *let $E[|X_n|] < \infty$ for each $n$. The sequence $\{X_n\}$ is called a martingale with respect to the Borel fields $\{\mathcal{B}_n\}$ if*

$$X_m = E[X_n|\mathcal{B}_m], \quad m < n, \tag{1}$$

*almost surely*. Think of a sequence of gambles at the times $n$. $\mathcal{B}_n$ can be thought of as corresponding to the information available to the player at time $n$. $X_n$ is the cumulative gain (or loss) of the player up to and including time $n$. Condition (1) then states that the sequence of plays is "fair" in the sense that the conditional net gain from time $m$ to $n$ given information up to and including time $m$ is zero. If the condition (1) is replaced by

$$X_m \geq E[X_n|\mathcal{B}_n], \quad m < n, \tag{2}$$

almost surely, the sequence $\{X_n\}$ is called a *supermartingale relative to $\{\mathcal{B}_n\}$*. Notice that if $Y_n$, $n = 1, 2, \ldots$, is a sequence of independent random variables with means equal to zero and $\mathcal{B}_n = (Y_j, j \leq n)$, the partial sums $X_n = \sum_{j=1}^{n} Y_j$ are a martingale relative to $\{\mathcal{B}_n\}$. If the means of the random variables $Y_n$ are all nonpositive, the sequence $(X_n)$ is a supermartingale relative to $(\mathcal{B}_n)$.

In Section c of Chapter V another example of a martingale had already been implicitly considered. Let $f$ be a bounded measurable function on a probability space with Borel field of events $\mathcal{B}$. The increasing family of subsigma-fields $C_n$ of $\mathcal{B}$ is assumed given with $C$ the sub-sigma-field of $\mathcal{B}$ that $\bigcup C_n$ generates. Set

$$X_n = E[f(w)|C_n]. \tag{3}$$

Then $(X_n)$ is a martingale relative to $(C_n)$ since

$$
\begin{aligned}
X_m &= E[f(w)|C_m] \\
&= E[E[f(w)|C_n]|C_m] = E[X_n|C_m]
\end{aligned} \tag{4}
$$

for $m < n$. In Section c of Chapter V we showed that $X_n = E[f(w)|C_n]$ converges almost everywhere to

$$X_\infty = E[f(w)|C], \tag{5}$$

as $n \to \infty$. This is an example of a martingale convergence theorem. A general result of this type will be derived in Section b. The following is a related example of a martingale. Consider a probability space with Borel field $\mathcal{B}$ and probability measure $P$ on $\mathcal{B}$. For each $n$ let $C_0^{(n)}$, $C_1^{(n)}$, . . . be a countable collection of disjoint measurable sets with union the whole space $\Omega$. Denote the Borel field generated by $C_j^{(n)}$ ($n$ fixed) by $C_n$. Further assume each $C_j^{(n+1)}$ is a subset of some $C_k^{(n)}$. Let $\nu$ be another measure on $\mathcal{B}$. Then if

$$X_n(w) = \frac{\nu(C_j^{(n)})}{P(C_j^{(n)})} \quad \text{for } w \epsilon C_j^{(n)}, \tag{6}$$

it can be easily verified that $X_m = E[X_n|C_m]$, $m < n$, so that $(X_n)$ is a martingale relative to $(C_n)$. A particular case of interest is that in which $\Omega = [0,1]$, $\mathcal{B}$ is the Borel field of Borel sets and $C_n$ is the finite field generated by the binary subintervals $\{x | j/2^n \le x < (j+1)/2^n\}$, $j = 0, 1, \ldots, 2^{n-1}$. Think of $P$ as the uniform measure on $[0,1]$ and $\nu$ as any other probability measure on the Borel sets. The function $X_n(w)$ is then the quotient of $\nu$ relative to the binary intervals of length $1/2^n$.

Consider a discrete time parameter Markov process with stationary transition mechanism. Let the $j$ step transition function be $P(j,x,A)$ with $x$, $A$ a point and a measurable subset of the state space. Still another illustration will be furnished in this context. Let $N_C$ be the *first hitting time* of the measurable subset $C$ (of the state space) for the Markov process $(Y_n, n = 0, 1, \ldots)$, that is

$$
\begin{aligned}
\{w|N_C(w) = k\} &= \{w|Y_j(w) \notin C, j = 0, \ldots, k-1, \ Y_k(w) \epsilon C\} \\
& \hspace{4cm} k = 0, 1, 2, \ldots \\
\{w|N_C(w) = \infty\} &= \{w|Y_j(w) \notin C, j = 0, 1, 2, \ldots\}. \tag{7}
\end{aligned}
$$

Formally, $N_C$ is an "improper" random variable in that we're possibly allowing it to take on the value $\infty$ with positive probability. However, it is easily seen that this will cause no difficulties in the following discussion. The random variable $N_C$ is an example of what is called a *stopping time relative to the increasing family of Borel fields* $\mathcal{B}_n = \mathcal{B}(Y_j, j \leq n)$, that is

$$\{w \mid N_C(w) = n\} \epsilon \mathcal{B}_n. \tag{8}$$

In the case of $N_C$ this also holds for $n = \infty$ with $\mathcal{B}_\infty = \mathcal{B}(Y_j, j = 0, 1, \ldots)$. Generally, *a stopping time can be defined with respect to any increasing family of Borel fields*. Let

$$h(y) = P[N_C(w) < \infty \mid Y_0(w) = y], \tag{9}$$

that is, $h(y)$ is the conditional probability of ever hitting $C$ given that one starts from $y$ initially. Set $X_n(w) = h(Y_n(w))$. Our claim is that $(X_n, n = 0, 1, \ldots)$ is a supermartingale relative to the Borel fields $\mathcal{B}_n = \mathcal{B}(Y_j, j \leq n)$. This follows since

$$E[X_n(w) \mid \mathcal{B}_m] = E[X_n(w) \mid Y_m] \tag{10}$$

by the Markov character of $(Y_n)$ and

$$\begin{aligned}
E[X_n \mid Y_m] &= E[h(Y_n(w)) \mid Y_m] \\
&= \int P(n - m, Y_n, dz) h(z) \\
&= P[Y_j \epsilon C \text{ for some } j \geq n \mid Y_m] \\
&\leq P[Y_j \epsilon C \text{ for some } j \geq m \mid Y_m] \\
&= h(Y_m) = X_n, \quad n \geq m. \tag{11}
\end{aligned}$$

Notice that if $\{X_n\}$ is a martingale relative to the sequence of Borel fields $\mathcal{B}_n$, the corresponding sequence of martingale differences $\{Z_n\}$ with $Z_n = X_n - X_{n-1}$ satisfy

$$E(Z_n \mid \mathcal{B}_m) \equiv 0, \quad m < n, \tag{12}$$

since

$$E(Z_n \mid \mathcal{B}_m) = E(X_n - X_{n-1} \mid \mathcal{B}_m) = X_m - X_m = 0, \tag{13}$$

if $m < n$. A sequence of martingale differences arises naturally in the context of a prediction problem. Let $(Y_n, n = \ldots, -1, 0, 1, \ldots)$ be a strictly stationary process with finite second moment, $EY_n^2 < \infty$. Consider the best (possibly nonlinear) predictor $Y_n^*$ of $Y_n$ depending on $Y_j, j \leq n - 1$, in terms of minimizing the mean square error of prediction

$$E|Y_n - Y_n^*|^2. \tag{14}$$

Let $\mathcal{B}_n = \mathcal{B}(Y_j, j \leq n)$ be the Borel field generated by $Y_j, j \leq n$. Then $Y_n^*$ is a $\mathcal{B}_{n-1}$ measurable random variable. One can show that

$$Y_n^* = E(Y_n | \mathcal{B}_{n-1}) \tag{15}$$

by the argument used to obtain the result in Problem 11 of Chapter IV. The prediction error is

$$Z_n = Y_n - Y_n^*. \tag{16}$$

The process $(Z_n)$ is a martingale difference relative to the sequence of Borel fields $(\mathcal{B}_n)$ since

$$\begin{aligned}
E(Z_n | \mathcal{B}_m) &= E(Y_n - Y_n^* | \mathcal{B}_m) \\
&= E(Y_n | \mathcal{B}_m) - E(E(Y_n | \mathcal{B}_{n-1}) | \mathcal{B}_m) \\
&= E(Y_n | \mathcal{B}_m) - E(Y_n | \mathcal{B}_m) = 0,
\end{aligned} \tag{17}$$

if $m < n$.

There are two fairly obvious but interesting remarks one can make about supermartingales. First notice that if $\{X_n\}$ and $\{Y_n\}$ are supermartingales relative to the sequence of Borel fields $\{\mathcal{B}_n\}$ then $aX_n + bY_n$ is a supermartingale relative to $\{\mathcal{B}_n\}$ for $a, b \geq 0$. The second is that $\{\min(X_n, Y_n)\}$ is then a supermartingale relative to $\{\mathcal{B}_n\}$ since if $X_m \leq Y_m$, $m < n$,

$$\begin{aligned}
\min(X_m, Y_m) = X_m &\geq E[X_n | \mathcal{B}_m] \\
&\geq E[\min(X_n, Y_n) | \mathcal{B}_m].
\end{aligned} \tag{18}$$

$(X_n)$ is called a submartingale if $(-X_n)$ is a supermartingale. Clearly $(X_n)$ is a martingale if and only if it is both a super and submartingale.

## b. Optional Sampling and a Martingale Convergence Theorem

Let $T$ be a stopping time relative to the increasing sequence of Borel fields $\mathcal{B}_n$. Then $T$ is called a *bounded stopping time* if there is a finite number $N$ such that $|T| \leq N$ with probability one. Given a stopping time $T$ and a process $X = (X_n)$ on a probability space, we mean by *the process $X$ stopped at $T$* (denoted by $X^T$)

$$X_n^T = X_{\min(n, T)}. \tag{1}$$

Also given a measurable set or event $A$, let $I_A$ denote the indicator function of $A$, that is,

$$I_A(w) = \begin{cases} 1 & \text{if } w \epsilon A \\ 0 & \text{if } w \notin A. \end{cases} \tag{2}$$

We first show that *if $X = (X_n)$ is a martingale (supermartingale) relative to the increasing sequence of Borel fields $\mathcal{B}_n$, $n = 0, 1, \ldots$, and $T$ is a stopping time, that then the stopped process $X^T$ is a martingale(supermartingale).* Notice that

$$X_n^T = \sum_{j=0}^{n-1} X_j I_{\{T=j\}} + X_n I_{\{T \geq n\}}. \tag{3}$$

This implies that

$$\begin{aligned} E[X_{n+1}^T - X_n^T | \mathcal{B}_n] &= E[I_{\{T \geq n+1\}}(X_{n+1} - X_n) | \mathcal{B}_n] \\ &= I_{\{T \geq n+1\}} E[(X_{n+1} - X_n) | \mathcal{B}_n] \overset{(\leq)}{=} 0. \end{aligned} \tag{4}$$

The integrability of $X_n^T$ follows from

$$|X_n^T| \leq |X_0| + \cdots + |X_n|. \tag{5}$$

The proof is essentially complete.

The result just derived on a stopped martingale suggests the following theorem on optional sampling. Before stating the theorem a little additional notation is introduced. If $T$ is a stopping time, the Borel field $\mathcal{B}_T$ consists of all events $A$ such that

$$A \cap \{T \leq n\} \epsilon \mathcal{B}_n. \tag{6}$$

Also given the event $A$, let the stopping time $T_A$ be such that

$$T_A(w) = \begin{cases} T(w) & \text{if } w \epsilon A \\ +\infty & \text{if } w \notin A. \end{cases} \tag{7}$$

**Theorem (optional sampling):** *Let $X$ be a supermartingale (martingale) relative to the increasing family of Borel fields $\mathcal{B}_n$, $n = 0, 1, \ldots$, with $S$ and $T$ two bounded stopping times such that $S \leq T$. The random variables $X_S$ and $X_T$ are then integrable with*

$$X_S \overset{(=)}{\geq} E[X_T | \mathcal{B}_S] \tag{8}$$

*with probability one.*

Assume that $k$ is a large enough integer such that $k \geq T$ with probability one. Introduce the process $Y = (Y_n)$

$$Y_n = X_{\min(T,n)} - X_{\min(S,n)}. \qquad (9)$$

The details of the proof will be given for the case in which $X$ is a supermartingale since the result for $X$ a martingale follows easily from that. The process $Y$ will be shown to be a supermartingale. Notice that $Y_0 = 0$ and $Y_k = X_T - X_S$. One can write

$$Y_n = \sum_{j=1}^{n} X_{j-1}(I_{\{S < j-1 = T\}} - I_{\{S = j-1 < T\}}) + X_n I_{\{S < n \leq T\}}. \qquad (10)$$

This implies that

$$
\begin{aligned}
E[Y_{n+1} - Y_n | \mathcal{B}_n] &= E[I_{\{S < n+1 \leq T\}}(X_{n+1} - X_n) | \mathcal{B}_n] \\
&= I_{\{S < n+1 \leq T\}} E[X_{n+1} - X_n | \mathcal{B}_n] \leq 0, \qquad (11)
\end{aligned}
$$

so that $Y$ is a supermartingale. Since $Y$ is a supermartingale we must have $0 \geq E(Y_k) = E(X_T - X_S)$. To obtain (8) consider any event $A \epsilon \mathcal{B}_S$ and replace $S$ and $T$ in the inequality just obtained by $S' = \min(S_A, k)$ and $T' = \min(T_A, k)$ respectively. The inequality

$$0 \geq E(X_{T'} - X_{S'}) = \int_A (X_T - X_s) \, dP \qquad (12)$$

follows. The definition of conditional expectation and inequality (12) imply (8).

As an immediate corollary the following result on the retention of the martingale property is obtained.

**Corollary:** *Let $X = (X_n)$ be a martingale relative to the sequence of Borel fields $\mathcal{B}_n$. Assume that $T_1 \leq T_2 \leq \cdots$ are a sequence of bounded stopping times. Then the derived optionally sampled process $(X_{T_n})$ is a martingale.*

There is an immediate interpretation of this corollary. Suppose that $X = (X_n)$ represents the results in a sequence of fair games. If the gambler chooses not to play every time but only at the stopping times $T_1 \leq T_2 \leq \cdots$ (so that he has no knowledge of the future), he is still playing a sequence of fair games.

We now state and derive an integral inequality for supermartingales. *Let $X = (X_n)$ be a supermartingale relative to the Borel fields $\mathcal{B}_n$, $n = 0, 1, \ldots$, and the event $A \epsilon \mathcal{B}_0$. Then if $c > 0$,*

$$cP(A \cap \{ \inf_{0 \leq k \leq m} X_k < -c \}) \leq \int_{A \cap \{ \inf_{0 \leq k \leq m} X_k < -c \}} X_j \qquad (13)$$

*for $j \geq m$.* First set

$$Y_n(w) = \begin{cases} 0 & \text{if } w \epsilon A \\ X_n(w) & \text{if } w \notin A. \end{cases} \tag{14}$$

The sequence $Y = (Y_n)$ is a supermartingale since

$$E(Y_{n+1} - Y_n | \mathcal{B}_n) = E(I_A(X_{n+1} - X_n) | \mathcal{B}_n)$$
$$= I_A E(X_{n+1} - X_n | \mathcal{B}_n) \leq 0. \tag{15}$$

Let

$$T(w) = \inf\{n | n \leq m, \, Y_n(w) \leq -c\}. \tag{16}$$

If no such integer $n$ exists set $T(w) = m$. The random variable $T$ is a stopping time bounded by $m$ so that the theorem on optional sampling can be applied. Thus $E[Y_T] \geq E[Y_m]$ and so

$$E[Y_m] \leq E[Y_T] \leq -cP[A \cap \{\inf_{n \leq m} X_n \leq -c\}]$$

$$+ \int_{(A \cap \{\inf_{n \leq m} X_n > -c\})} X_m, \tag{17}$$

or

$$cP[A \cap \{\inf_{n \leq m} X_n \leq -c\}] \leq \int_{A \cap \{\inf_{n \leq m} X_n \leq -c\}} X_m. \tag{18}$$

The inequality just derived can be used to prove a martingale convergence theorem.

**Theorem:** *Let $X = (X_n)$ be a martingale relative to the family of Borel fields $\mathcal{B}_n$, $n = 0, 1, \ldots$, with $E|X_j| \leq M < \infty$ for all $j$. Then*

$$\lim_{j \to \infty} X_j = X \tag{19}$$

*exists with probability one where $E|X| \leq M$.*

The proof is indirect. Assume there is no convergence with probability one. Then there are numbers $a < b$ such that the set

$$D = \{\overline{\lim_{k \to \infty}} \, X_k > b, \, \underline{\lim_{k \to \infty}} \, X_k < a\} \tag{20}$$

has $P(D) > 0$. Let

$$B_{mn} = \{\sup_{m \leq k \leq n} X_k > b\}$$
$$A_{mn} = \{\inf_{m \leq k \leq n} X_k < a\}. \tag{21}$$

We can then find a sequence of integers $n_1 < n_2 < \cdots$ such that

$$P(D \cap A_{1n_1}) \geq P(D)(1 - 4^{-1})$$
$$P(D \cap A_{1n_1} \cap B_{n_1 n_2}) \geq P(D)(1 - 4^{-1} - 4^{-2})$$
$$P(D \cap A_{1n_1} \cap B_{n_1 n_2} \cap A_{n_2 n_3}) \geq P(D)(1 - 4^{-1} - 4^{-2} - 4^{-3})$$
$$\cdots$$

(22)

Let $D_1 = A_{1n_1}$, $D_2 = A_{1n_1} \cap B_{n_1 n_2}$, $\ldots$ Since $X$ is a martingale, $X$ and $-X$ are supermartingales. The inequality (18) implies that

$$\int_{D_k} X_j \geq b P(D_k),$$

(23)

or

$$\int_{D_k} X_j \leq a P(D_k),$$

according as $k$ is even or odd, for $j > n_k$. This implies that

$$\int_{D_k - D_{k+1}} X_j \geq (b - a) P(D_{k+1}) \geq \tfrac{2}{3}(b - a) P(D),$$

(24)

if $j > n_{k+1}$ ($k$ even). Then

$$E|X_j| \geq \tfrac{2}{3} r(b - a) P(D)$$

(25)

with $r$ any given positive integer if $j$ is sufficiently large. But this contradicts the boundedness of $E|X_j|$ uniformly in $j$.

We shall say that a martingale $X = (X_n)$ with respect to $\mathcal{B}_n$, $n = 1, 2, \ldots$, is *closed* if there is a random variable $Y$, $E|Y| < \infty$, such that

$$X_n = E[Y | \mathcal{B}_n]$$

(26)

with probability one. The following result indicates when a martingale is closed.

**Corollary:** *A martingale $X = (X_n)$ is closed if and only if it is uniformly integrable.*

Assume $X = (X_n)$ is closed by $Y$. Then

$$\left| \int_{\{|X_n| > c\}} X_n \right| = \left| \int_{\{|X_n| > c\}} Y \right| \leq \int_{\{|X_n| > c\}} |Y|,$$

(27)

and

$$P\{|X_n| > c\} \leq \frac{1}{c} E|X_n| \leq \frac{1}{c} E|Y|,$$

(28)

so that the first integral on the left of (27) is less than $\varepsilon$ for $c$ sufficiently large. Thus $X$ is uniformly integrable. Conversely, assume that $(X_n)$ is uniformly integrable. Then $E|X_n| \leq M < \infty$ for some constant $M$. The martingale convergence theorem implies that there is a random variable $Y = \lim_{n \to \infty} X_n$ with $E|Y| < \infty$. The uniform integrability implies that

$$E|X_n - Y| \to 0, \tag{29}$$

as $n \to \infty$. Since

$$\int_A X_m = \int_A Y = \int_A \lim_{n \to \infty} X_n, \tag{30}$$

for $A \epsilon \mathcal{B}_s$, $s \leq m$, $Y$ closes $X$.

As an application of these results, consider the illustration of section $a$ in which one had a probability space with Borel field $\mathcal{B}$, measures $\nu$ and $P$ on $\mathcal{B}$ and the increasing family of subfields $\mathcal{C}_j$. The sequence $X_n(w)$ as given by (a.6) is a martingale with $EX_n(w) = E|X_n(w)| \leq \nu(\Omega) < \infty$. The martingale convergence theorem of this section implies that

$$\lim_{n \to \infty} X_n(w) = X_\infty(w) \tag{31}$$

exists almost surely with $EX_\infty(w) \leq \nu(\Omega)$. If $\nu$ is absolutely continuous with respect to $P$ so that

$$\nu(B) = \int_B f(w)\, dP \tag{32}$$

the martingale is closed since

$$X_n = E[f(w)|\mathcal{C}_n]. \tag{33}$$

The corollary then implies that

$$E|X_n - X_\infty| \to 0, \tag{34}$$

as $n \to \infty$. Further, if $\mathcal{B} = \bigvee_{j=1}^{\infty} \mathcal{C}_j$ ($\mathcal{B}$ is the smallest Borel field containing $\bigcup_{j=1}^{\infty} \mathcal{C}_j$), we must have

$$f(w) = X_\infty(w) \tag{35}$$

almost surely.

## c. A Central Limit Theorem for Martingale Differences

The principal result of this section is concerned with a stationary sequence of martingale differences $X = (X_n)$ relative to a sequence of Borel fields $\mathcal{B}_n$, $n = \ldots, -1, 0, 1, \ldots$, that is,

$$E(X_n|\mathcal{B}_m) \equiv 0, \quad m < n. \tag{1}$$

**Theorem:** *Let $X = (X_n)$ be a stationary ergodic sequence of martingale differences with finite second moment $EX_n^2 = 1$. Then*

$$\frac{1}{\sqrt{n}} \sum_{j=1}^{n} X_j \tag{2}$$

*is asymptotically normally distributed with mean zero and variance one as $n \to \infty$.*

Results of this type are initially due to P. Lévy. The particular result stated is due to Billingsley. The format of the proof we give is modeled on one of Brown and Eagleson [A2]. The following lemma is helpful in the proof.

**Lemma:** *Let $X_n$, $w_n$, $n = 1, 2, \ldots$, be random variables with $\phi(t)$ a characteristic function. Assume for some fixed $t$ that $w_n^{-1}$ converges to $\phi(t)^{-1}$ in mean of order one and that*

$$\lim_{n \to \infty} E(w_n^{-1} \cdot \exp(itX_n) - 1) = 0. \tag{3}$$

*Then*

$$\lim_{n \to \infty} E \exp(itX_n) = \phi(t). \tag{4}$$

The lemma follows since

$$
\begin{aligned}
|E(\exp(itX_n) - \phi(t))| &\leq |E \exp(itX_n)(1 - w_n^{-1}\phi(t))| \\
&\quad + |E\phi(t)(w_n^{-1}\exp(itX_n) - 1)| \\
&\leq E|1 - w_n^{-1}\phi(t)| \\
&\quad + |\phi(t)| \, |E(w_n^{-1}\exp(itX_n) - 1)| \\
&\leq E|w_n^{-1} - \phi(t)^{-1}| \\
&\quad + |E(w_n^{-1}\exp(itX_n) - 1)| \\
&\to 0 \quad \text{as } n \to \infty. \tag{5}
\end{aligned}
$$

We now go on to prove the theorem. Let

$$S_n = \frac{1}{\sqrt{n}} \sum_{j=1}^{n} X_j \tag{6}$$

$$\sigma_{nk}^2 = \frac{1}{n} E(X_k^2 | \mathcal{B}_{k-1})$$

$$V_{nk}^2 = \sum_{j=1}^{k} \sigma_{nj}^2, \qquad b_n = \max_{1 \le j \le n} \sigma_{nj}^2.$$

Because of the ergodicity of the process $X = (X_n)$ and $EX_n^2 = 1$, it follows that $V_{nn}^2 \to EX_1^2 = 1$ with probability one as $n \to \infty$. Thus

$$\lim_{n \to \infty} P(V_{nn}^2 \le 2) = 1. \tag{7}$$

It is clear that

$$E(X_1^2 I_{\{|X_1| > \varepsilon \sqrt{n}\}}) \to 0, \tag{8}$$

as $n \to \infty$ where $\varepsilon$ is any fixed positive number. But this implies that

$$\frac{1}{n} \sum_{j=1}^{n} E(X_j^2 I_{\{|X_j| \ge \varepsilon \sqrt{n}\}} | \mathcal{B}_{j-1}) \to 0 \tag{9}$$

in probability as $n \to \infty$ since (7) is the expected value of the non-negative random variable (8). Notice that (9) implies that

$$b_n \to 0 \tag{10}$$

in probability as $n \to \infty$ since

$$b_n \le \varepsilon^2 + \frac{1}{n} \max_j E(X_j^2 I_{\{|X_j| \ge \varepsilon \sqrt{n}\}} | \mathcal{B}_{j-1})$$

$$\le \varepsilon + \frac{1}{n} \sum_{j=1}^{n} E(X_j^2 I_{\{|X_j| \ge \varepsilon \sqrt{n}\}} | \mathcal{B}_{j-1}). \tag{11}$$

Let

$$X_{nk}^* = X_k I_{\{V_{nk}^2 \le 2\}}, \quad k = 1, 2, \ldots \tag{12}$$

Then $X_{n_1}^*, \ldots, X_{nn}^*$ is a martingale difference sequence relative to $\mathcal{B}_1, \ldots, \mathcal{B}_n$ since $I_{\{V_{nk}^2 \le 2\}}$ is $\mathcal{B}_{k-1}$ measurable. If we set

$$V_{nk}^{*2} = \frac{1}{n} \sum_{i=1}^{k} E(X_{ni}^{*2}|\mathcal{B}_{i-1})$$

then

$$V_{nn}^{*2} = \frac{1}{n} \sum_{i=1}^{n} I_{(V_{ni}^2 \leq 2)} E(X_i^2|\mathcal{B}_{i-1}) \leq 2 \tag{13}$$

with probability one. Now

$$\lim_{n \to \infty} P\left[\bigcap_{j=1}^{n} (X_{nj}^* = X_j)\right] = 1, \tag{14}$$

since $(X_{nj}^* \neq X_j) = (V_{nj}^2 > 2)$ and $(V_{nj}^2 > 2) \subset (V_{nn}^2 > 2)$ so that $P[\bigcup_{j=1}^{n} (X_j \neq X_{nj}^*)] \subset P(V_{nn}^2 > 2)$. Thus, it is enough to carry out the proof with $X_{nj}^*$'s instead of $X_j$'s. However, the proof follows just as if we took the $X_j$'s and assumed $V_{nn}^2 \leq 2$ with probability one and this is the way we shall continue. Let $g_t(x) = (e^{itx} - 1 - itx)x^{-2}$. Since

$$|g_t(x)x^2| \leq \frac{1}{2} t^2 x^2, \tag{15}$$

formula (9) implies that if $X_{nk} = \frac{1}{\sqrt{n}} X_k$,

$$\sum_{k=1}^{n} E(g_t(X_{nk})X_{nk}^2|\mathcal{B}_{k-1}) \to -\frac{t^2}{2}$$

in probability as $n \to \infty$. Because of the boundedness of $V_{nn}^2$

$$\exp\left(\sum_{k=1}^{n} E(g_t(X_{nk})X_{nk}^2|\mathcal{B}_{k-1})\right) \to \exp\left(-\frac{t^2}{2}\right) \tag{16}$$

in mean of order one as $n \to \infty$. Let

$$F_n = \exp\left[itS_n - \sum_{k=1}^{n} E(g_t(X_{nk})X_{nk}^2|\mathcal{B}_{k-1})\right] - 1 = \sum_{k=1}^{n} G_{nk}, \tag{17}$$

where

$$G_{nk} = \exp\left[it\sum_{j=1}^{k-1} X_{nj} - \sum_{j=1}^{k} E(g_t(X_{nj})X_{nj}^2|\mathcal{B}_{j-1})\right]$$
$$\times \{\exp(itX_{nk}) - \exp(E(g_t(X_{nk})X_{nk}^2|\mathcal{B}_{k-1}))\}. \tag{18}$$

Then

$$E(G_{nk}|\mathcal{B}_{k-1}) = -\{Q(E(g_t(X_{nk})X_{nk}^2|\mathcal{B}_{k-1}))\}$$

$$\exp\left[it\sum_{j=1}^{k-1} X_{nj} - \sum_{j=1}^{k} E(g_t(X_{nj})X_{nj}^2|\mathcal{B}_{j-1})\right], \qquad (19)$$

with $Q(x) = e^x - 1 - x$. Now

$$Q(x) \leq \tfrac{1}{2}|x|^2 e^{|x|}. \qquad (20)$$

This implies that

$$|E(G_{nk}|\mathcal{B}_{k-1})| \leq \tfrac{1}{4}t^4\sigma_{nk}^4 \exp(2t^2). \qquad (21)$$

Thus

$$|EF_n| \leq E\sum_{k=1}^{n} |E(G_{nk}|\mathcal{B}_{k-1})|$$

$$\leq \tfrac{1}{4}t^4 \exp(2t^2)E\sum_{k=1}^{n}\sigma_{nk}^4$$

$$\leq \tfrac{1}{4}t^4 \exp(2t^2)E(b_n V_{nn}^2) \to 0 \qquad (22)$$

as $n \to \infty$. Now use the lemma with $X_n = S_n$, $\phi(t) = \exp(-t^2/2)$ and

$$w_n = \exp\left(\sum_{k=1}^{n} E(g_t(X_{nk})X_{nk}^2|\mathcal{B}_{k-1})\right), \qquad (23)$$

and we find that

$$\lim_{n\to\infty} E\exp(itS_n) = e^{-t^2/2} \qquad (24)$$

for every $t$. The desired result follows by using the continuity theorem for characteristic functions.

In the following corollary we have a central limit theorem for a broader class of stationary processes. The basic idea is that of introducing an appropriately defined martingale and applying the foregoing theorem. This technique was first used by Gordin [A6].

**Corollary**: *Let $Y = (Y_n)$, $n = \cdots, -1, 0, 1, \cdots$ be a strictly stationary process with $EY_j \equiv 0$, $EY_j^2 < \infty$. If $\mathcal{B}_n = (Y_k, k \leq n)$ is the Borel field generated by $Y_k, k \leq n$, assume that*

$$\sum_{j=1}^{\infty}\{E(E(Y_0|\mathcal{B}_{nj}))^2\}^{1/2} < \infty, \qquad (25)$$

*and*

$$EY_0^2 + 2 \sum_{k=1}^{\infty} E(Y_k Y_0) = \sigma^2 > 0. \tag{26}$$

*Then*

$$\frac{1}{\sqrt{n}\,\sigma} \sum_{j=1}^{n} Y_j \tag{27}$$

*is asymptotically normal with mean zero and variance one.*

Let $S_n = \sum_{j=1}^{n} Y_j$. Then

$$ES_n^2 = nEY_0^2 + 2 \sum_{k=1}^{n} (n-k) E(Y_k Y_0). \tag{28}$$

Now

$$|E(Y_k Y_0)| = |E(E(Y_k|\mathcal{B}_0)Y_0)| \leq \{E(Y_0^2)E(E(Y_k|\mathcal{B}_0)^2)\}^{1/2}. \tag{29}$$

Notice that (25), (26), (28) and (29) imply that

$$\frac{1}{n} ES_n^2 \to \sigma^2 > 0, \tag{30}$$

as $n \to \infty$. Let

$$u_r = E(Y_r|\mathcal{B}_0) - E(Y_r|\mathcal{B}_{-1}), \quad 0 \leq r < \infty. \tag{31}$$

The inequality (25) implies that $\{E|\sum_{r=0}^{\infty} u_r|^2\}^{1/2} \leq \sum_{r=0}^{\infty} \{E|u_r|^2\}^{1/2} < \infty$ so that we can set

$$X_0 = \sum_{r=0}^{\infty} u_r,$$

$$X_r = \tau^r X_0, \tag{32}$$

where $\tau$ is the shift operator. Notice that $X = (X_r)$ is a process of martingale differences since

$$E(X_r|\mathcal{B}_{r-1}) \equiv 0, \tag{33}$$

almost surely. The basic properties of conditional expectations imply that

$$
\begin{aligned}
E(u_r u_{r+k}) &= E(E(Y_r|\mathcal{B}_0)E(Y_{r+k}|\mathcal{B}_0)) \\
&\quad - E(E(Y_r|\mathcal{B}_{-1})E(Y_{r+k}|\mathcal{B}_{-1})) \\
&= E(Y_{r+k}E(Y_r|\mathcal{B}_0)) - E(Y_{r+k}E(Y_r|\mathcal{B}_{-1})) \\
&= E(Y_k E(Y_0|\mathcal{B}_{-r})) - E(Y_k E(Y_0|\mathcal{B}_{-r-1})). \quad (34)
\end{aligned}
$$

Now to evaluate

$$
\begin{aligned}
E|X_0|^2 &= \lim_{n\to\infty} E\Big|\sum_{r=0}^n u_r\Big|^2 \\
&= \lim_{n\to\infty}\left[\sum_{r=0}^n Eu_r^2 + 2\sum_{k=1}^n \sum_{r=0}^{n-k} E(u_r u_{r+k})\right] \\
&= \lim_{n\to\infty}\big[EY_0^2 - E(Y_0 E(Y_0|\mathcal{B}_{-n-1})) \\
&\qquad + 2\sum_{k=1}^n \{E(Y_k Y_0) - E(Y_k E(Y_0|\mathcal{B}_{-n+k-1}))\}\big] \\
&= EY_0^2 + 2\sum_{k=1}^\infty E(Y_k Y_0) = \sigma^2 > 0. \quad (35)
\end{aligned}
$$

Set $T_n = \sum_{j=1}^n X_j$. Since $X$ is a martingale, by the preceding theorem

$$
\frac{1}{\sqrt{n}\,\sigma} T_n \quad (36)
$$

is asymptotically normal with mean zero and variance one. Clearly $ET_n^2 = n\sigma^2$. Thus, if one can show that $\frac{1}{n} ES_n T_n \to \sigma^2$ as $n \to \infty$, it will imply that $\frac{1}{n} E(S_n - T_n)^2 \to 0$ as $n \to \infty$ and consequently the desired conclusion of the corollary. Again using properties of conditional expectations we find

$$
\begin{aligned}
ES_n T_n &= \sum_{i=1}^n \sum_{j=1}^n E(Y_i X_j) \\
&= nEX_0 Y_0 + \sum_{j=1}^n (n-j)\{E(Y_0 X_j) + E(Y_0 X_{-j})\} \\
&= \sum_{j=0}^{n-1} (n-j)E(Y_0 X_{-j}) \\
&= \sum_{j=0}^{n-1}(n-j)E\Big\{Y_0 \sum_{r=0}^\infty (E(Y_{-j+r}|\mathcal{B}_{-j}) - E(Y_{-j+r}|\mathcal{B}_{-j-1}))\Big\}
\end{aligned}
$$

$$= \left( \sum_{j=0}^{n-1} \sum_{r=0}^{\infty} (n-j) - \sum_{j=1}^{n} \sum_{r=1}^{\infty} (n-j+1) \right) E(Y_0 E(Y_{-j+r}|\mathcal{B}_{-j}))$$

$$= \sum_{j=0}^{n-1} (n-j) E(Y_0 Y_{-j}) - \sum_{j=1}^{n} \sum_{r=1}^{\infty} E(Y_0 E(Y_{-j+r}|\mathcal{B}_{-j}))$$

$$+ n \sum_{r=1}^{\infty} E(Y_0 E(Y_r|\mathcal{B}_0))$$

$$= \sum_{j=0}^{n-1} (n-j) E(Y_0 Y_j) - \sum_{j=1}^{n} \sum_{r=1}^{\infty} E(Y_j E(Y_r|\mathcal{B}_0))$$

$$+ n \sum_{r=1}^{\infty} E(Y_0 Y_r). \tag{37}$$

Thus $n^{-1} E((S_n T_n) \to \sigma^2$ as $n \to \infty$.

## d. Problems

**1.** Let $X_1, X_2, \ldots$ be a Markov chain with transition probability matrix $P$. Assume that $A$ is a set of recurrent states of the chain. Let $N_1 < N_2 < \cdots$ be the successive times that the chain takes values in $A$. Show that $X_{N_1}, X_{N_2}, \ldots$ is a Markov chain.

**2.** Consider $X_1, X_2, \ldots$ a sequence of independent identically distributed random variables with $E|X_j| < \infty$ and $N$ a stopping time on the sample space with $EN < \infty$. Show that

$$E(X_1 + \cdots + X_N) = E(N)E(X).$$

(This result is commonly referred to as Wald's equation.)

**3.** Let $X_1, X_2, \ldots$ be a Markov chain with $N_1 < N_2 < \cdots$ a sequence of stopping times. Show that $X_{N_1}, X_{N_2}, \ldots$ is a Markov chain.

**4.** Let $P$ be the transition probability function of a Markov chain. The vector $\mathbf{h}$ is said to be subharmonic (harmonic) with respect to $P$ if $\mathbf{h} \overset{(=)}{\geq} P\mathbf{h}$. If $p_{i,i+1} = p$, $p_{i,i-1} = q$ with $0 < p < 1$, $q = 1 - p$, $i = 0, \pm 1, \ldots$, determine the harmonic and superharmonic functions relative to $P$.

**5.** Let $X_1, X_2, \ldots$ be a Markov chain with transition probability matrix $P$. Suppose that $h$ is superharmonic relative to $P$ and that

$E|h(X_k)| < \infty$ for all $k$. Show that $Y_k = h(X_k), k = 1, 2, \ldots$, is a supermartingale.

**6.** Consider the random walk on the lattice points $(i,j)$, $i,j = 0, \pm 1, \ldots$ in the plane with

$$p_{(i,j),(i+1,j)} = p_{(i,j),(i-1,j)} = p_{(i,j),(i,j+1)} = p_{(i,j),(i,j-1)} = \tfrac{1}{4}.$$

Let $A$ be a set of lattice points with $B$ a subset of $A$. Let $N$ be the first hitting time of the set $A$ where it is understood that $N = 0$ if one starts in $A$. Consider.

$$P[X_N \epsilon B \subset A | X_0 = (i,j)] = h(i,j),$$

where $X_1, X_2, \ldots$ is the random walk. Show that

$$h(i,j) = \begin{cases} 1 & \text{if } (i,j)\epsilon B \\ 0 & \text{if } (i,j)\epsilon A - B \end{cases}$$

and that

$$h(i,j) = \tfrac{1}{4}\{h(i,j+1) + h(i,j-1) + h(i+1,j) + h(i-1,j)\}$$
if $(i,j)\notin A$.

**7.** In the context of problem 6 let

$$P[X_N \epsilon B \subset A, N = n | X_0 = (i,j)] = h_n(i,j),$$

where $N$ is the first hitting time of the set $A$. Show that

$$h_0(i,j) = \begin{cases} 1 & \text{if } (i,j)\epsilon B \\ 0 & \text{if } (i,j)\notin B \end{cases}$$

while

$$h_n(i,j) = 0 \quad \text{if } (i,j)\epsilon A \text{ and } n > 0$$

and

$$h_n(i,j) = \tfrac{1}{4}(h_{n-1}(i+1,j) + h_{n-1}(i-1,j) + h_{n-1}(i,j+1)$$
$$+ h_{n-1}(i,j-1))$$

if $(i,j)\notin A$, $n > 0$. Also

$$h(i,j) = \sum_{n=0}^{\infty} h_n(i,j).$$

**8.** Let $(X_n)$ be a martingale difference sequence relative to the sequence of Borel fields $\mathcal{B}_n$, $n = 1, 2, \ldots$ . Set

$$S_n = \sum_{j=1}^{n} X_j, \quad \sigma_n^2 = E(X_n^2 | \mathcal{B}_{n-1}),$$

$$V_n^2 = \sum_{j=1}^{n} \sigma_j^2, \quad \text{and} \quad s_n^2 = ES_n^2 = EV_n^2.$$

Show that if $s_n^2 \uparrow \infty$

$$s_n^{-2} \max_{j \le n} \sigma_j^2 \to 0 \quad \text{in probability}$$

$$V_n^2 \to \sigma^2 \text{ (constant)} \quad \text{in probability}$$

$$s_n^{-2} \sum_{j=1}^{n} E(X_j^2 I(|X_j| \ge \varepsilon s_n) | \mathcal{B}_{n-1}) \to 0 \quad \text{in probability}$$

as $n \to \infty$ for each $\varepsilon > 0$, then $S_n/s_n$ converges in distribution as $n \to \infty$ to a normal distribution with mean 0 and variance $\sigma^2$.

# Notes

1. In the case of a continuous time parameter $t$, the Borel fields $_t$ are assumed to be nondecreasing and right continuous in the sense that

$$_t = _{t+} = \bigcap_{s > t} {}_s$$

(see Meyer [A10]).

2. P. Lévy (see his *Theorié de l'Addition des Variables Aléatoires*, 1937) obtained the earliest results on martingales. Much of the development of interest in martingales and their application is due to the research and influence of J. L. Doob [12].

3. A broader class of limit theorems for martingales can be found in the paper of Brown and Eagleson [A2].

# IX

# ADDITIONAL TOPICS

## a. A Zero-One Law

Let $A_1, A_2, \ldots$ be a sequence of events of the probability space of interest and therefore by implication elements of the sigma-field on which the probability measure is defined. Our object is to compute the probability of the set of points belonging to an infinite number of the sets $A_i, i = 1, 2, \ldots$ . Loosely speaking, we are interested in the probability that the $A_i$ occur infinitely often. For convenience, this set will be referred to as $A_i$ i.o. and its complement (the $A_i$ occur finitely often) as $A_i$ f.o. The event $A_i$ i.o. belongs to the sigma-field of sets on which the probability is defined since

$$A_i \text{ i.o.} = \bigcap_{n=1}^{\infty} \bigcup_{i=n}^{\infty} A_i. \tag{1}$$

Therefore

$$P(A_i \text{ i.o.}) \leq P(\bigcup_{i=n}^{\infty} A_i) \leq \sum_{i=n}^{\infty} P(A_i) \tag{2}$$

so that *if*

$$P(A_i \text{ i.o.}) > 0 \tag{3}$$

*then*

$$\sum_{1}^{\infty} P(A_i) \tag{4}$$

*diverges.*

A parallel statement in the opposite direction is not possible for general $A_i$. However, it is feasible for independent $A_i$. Let us therefore now assume that the $A_i$ are a sequence of independent events. Now

$$A_i \text{ f.o.} = (\overline{A_i \text{ i.o.}}) = \bigcup_{n=1}^{\infty} \bigcap_{i=n}^{\infty} \bar{A}_i \tag{5}$$

(see Problem 1, Chapter II). Since the $A_i$ are independent

$$P(\bigcap_{i=n}^{\infty} \bar{A}_i) = \prod_{i=n}^{\infty} (1 - P(A_i)) \leq \exp(-\sum_{i=n}^{\infty} P(A_i)). \tag{6}$$

**200**

If $\Sigma P(A_i)$ diverges, $P(\bigcap_{j=1}^{\infty} \bar{A}_i) = 0$ by (6) and thus

$$P(A_i \text{ f.o.}) = 0. \tag{7}$$

It therefore follows that *if $A_i$, $i = 1, 2, \ldots$, is a sequence of independent events, $P(A_i \text{ i.o.})$ can only assume the two values 0 and 1. The values 0 and 1 correspond to the convergence and divergence of $\Sigma P(A_i)$ respectively.* Such a result is an example of a zero-one law.

# b. Markov Chains and Independent Random Variables

Let $X_n$, $n = 0, \pm 1, \ldots$, be a finite state stationary Markov chain with transition probability matrix

$$\mathbf{P} = (p_{i,j}; i, j = 1, \ldots, m)$$
$$p_{i,j} = P(X_{n+1} = j | X_n = i) \tag{1}$$

and instantaneous invariant probability vector

$$\mathbf{p} = (p_i; i = 1, 2, \ldots, m)$$
$$p_i = P(X_n = i) \tag{2}$$
$$\mathbf{p}\mathbf{P} = \mathbf{p}.$$

For convenience we *assume* that *the transition probabilities $p_{i,j}$ are all positive.* This implies that the process $X_n$ is an ergodic *process without any periodic states* (see Problem 11, Chapter III). Our object is to show that *the process $X_n$ can be represented as a one-sided function $f$ of independent uniformly distributed* (on $[0,1]$) *random variables $\xi_n$ and its shifts, that is,*

$$X_n = f(\xi_n, \xi_{n-1}, \ldots). \tag{3}$$

To carry out the construction, a continuous state stationary Markov process $\{Y_n\}$, $n = 0, \pm 1, \ldots$, is introduced. Let

$$Y_n = X_n - U_n \qquad n = 0, \pm 1, \ldots \tag{4}$$

where the $U_n$ are a sequence of independent uniformly distributed random variables $[0,1]$, and the process $\{U_n\}$ is independent of the process $\{X_n\}$. Notice that we have enlarged the probability space of the $X_n$ process by adjoining the $U_n$ process. The process $Y_n$ is a stationary ergodic Markov process with transition probability density

$$p(y|y') = p_{i,j} \text{ if } i - 1 \le y' < i \text{ and } j - 1 \le y < j \tag{5}$$

and instantaneous density function

$$p(y) = p_j \text{ if } j - 1 \le y < j, \tag{6}$$

$i, j = 1, 2, \ldots, m$. Notice that

$$X_n = [Y_n] + 1 \tag{7}$$

where $[y]$ is the greatest integer less than $y$. Let $F(y|y')$ be the conditional distribution function of $Y_n$ given $Y_{n-1}$. Thus

$$F(y|y') = \sum_{k=1}^{j-1} p_{i,k} + p_{i,j}(y - j + 1) \tag{8}$$

if $i - 1 \le y' < i$ and $j - 1 \le y < j$. Notice that $F(y|y')$ with $y'$ fixed, $0 \le y' < m$, increases strictly from zero to one as $y$ ranges from zero to $m$. Further, the function $z = F(y|y')$ with $y$ fixed, $0 \le y < m$, maps each $y'$ interval of the form $i - 1 \le y' < i, i = 1, \ldots, m$, into a single $z$ value. Of course, the $z$ value depends on the fixed values of $y$ and $i$. Introduce the random variable

$$\xi_n = F(Y_n|Y_{n-1}). \tag{9}$$

The random variable $\xi_n = F(Y_n|Y_{n-1})$ is uniformly distributed on $[0,1]$ and independent of $Y_{n-1}$ (see Problem IV.5). The Markovian property of the $Y_n$ sequence implies that $\xi_n$ is independent not only of $Y_{n-1}$ but also of all the preceding random variables $Y_{n-2}, Y_{n-3}, \ldots$. The $\xi_n$ are therefore a sequence of independent uniformly distributed random variables on $[0,1]$.

Let

$$\delta = \min_i p_{i,1} > 0. \tag{10}$$

From (8) and (9) it follows that $0 < \xi_{n-1} < \delta$ implies that $0 \le Y_{n-1} < 1$. If it is also true that $0 < \xi_n < \delta$, then by (8)

$$Y_n = \xi_n/p_{1,1}. \tag{11}$$

Further, knowledge of $Y_n$ and $\xi_{n+1}$ determine $Y_{n+1}$ since

$$Y_{n+1} = F^{-1}(\xi_{n+1}|Y_n). \tag{12}$$

Here $F^{-1}(z|y') = y$ is the inverse function of $z = F(y|y')$ as a function of $y, 0 \le y < m$, with $y'$ fixed. It is now possible to see how the process $Y_n$ can be constructed out of the $\xi_n$ sequence by a one-sided function and its shifts. Let $A_n$ be the event

$$A_n = \{0 \le \xi_{n-1}, \xi_n < \delta\}. \tag{13}$$

Set

$$G_1(z,y) = F^{-1}(z|y) \tag{14}$$
$$G_2(z_1,z_2,y) = F^{-1}(z_1|G_1(z_2,y)) \tag{15}$$

and recursively

$$G_k(z_1, \ldots ,z_k,y) = F^{-1}(z_1|G_{k-1}(z_2, \ldots ,z_k,y)). \tag{16}$$

Consider the one-sided infinite sequence of independent random variables

$$\ldots , \xi_{n-2}, \xi_{n-1}, \xi_n \tag{17}$$

and the independent events

$$\ldots , A_{n-6}, A_{n-4}, A_{n-2} \tag{18}$$

defined in terms of these random variables. Notice that $\sum_{k=1}^{\infty} P(A_{n-2k})$ diverges since all the events have the same positive probability. By the zero-one law almost every realization of the sequence (17) must lie in one of the events $A_{n-2k}$, $k = 1, 2, \ldots$ . Suppose $j$ is the smallest positive integer for which the realization of (17) falls into an event $A_{n-2k}$ ($j$ of course depends on the realization). Then

$$Y_n = G_{2j}(\xi_n, \ldots , \xi_{n-2j+1}, \xi_{n-2j}/p_{1,1}) \tag{19}$$

with probability one. Since $X_n$ is given in terms of $Y_n$ by (7), $X_n$ is determined in terms of $\xi_n, \xi_{n-1}, \ldots$ .

Actually a necessary and sufficient condition that a general denumerable state stationary Markov chain have a representation of the form (3) is that the chain be ergodic and have no periodic states (see [67], [69]). The argument is more elaborate and the representation is not as simple as (19).

# c. A Representation for a Class of Random Processes

Let $X_t = X_t(w)$, $0 \leq t \leq T < \infty$, be a real-valued random process continuous in mean square. Further, we shall assume that our representation of the process is such that $X_t(w)$ is jointly measurable in $t$ and $w$. As before the mean value $EX_t(w)$ is taken to be identically zero for convenience. The mean square continuity of the process implies that the covariance function

$$r(t,\tau) = EX_t(w)X_\tau(w) \tag{1}$$

is continuous in $t$ and $\tau$ jointly on the square $0 \leq t, \tau \leq T$.

The representation of the process $X_t$ will be given in terms of the solutions $\varphi(t)$ of the integral equation

$$\varphi(t) = \lambda \int_0^T r(t,\tau)\varphi(\tau) \, d\tau \tag{2}$$

with kernel $r(t,\tau)$. A solution $\varphi(t)$ of equation (2) is called *an eigenfunction of the equation* and the corresponding number $\lambda$ *its eigenvalue*. Of course, we are only interested in nontrivial solutions $\varphi$ and hence the corresponding eigenvalue $\lambda \neq 0$. Further, the eigenfunctions of interest are to be decent in that they are square integrable, that is,

$$\int_0^T |\varphi(t)|^2 \, dt < \infty. \tag{3}$$

More can actually be said. The kernel $r(t,\tau)$ is uniformly continuous since it is continuous on the closed interval $0 \leq t, \tau \leq T$. This coupled with an application of the Schwarz inequality implies that the eigenfunctions $\varphi(t)$ of the integral equation (2) are all continuous. It has already been remarked that the covariance function of a random process is positive definite. Thus

$$\int_0^T |\varphi(t)|^2 \, dt = \lambda \iint_0^T \overline{\varphi(t)} r(t,\tau)\varphi(\tau) \, dt \, d\tau > 0. \tag{4}$$

It follows that *all eigenvalues of the integral equation are positive. The eigenfunctions $\varphi(t)$ can be taken as real-valued functions.* For even if $\varphi(t)$ were complex, by (2) its real and imaginary parts separately would be eigenfunctions with eigenvalue $\lambda$. Consider now any two eigenfunctions $\varphi_1(t)$, $\varphi_2(t)$ corresponding to distinct eigenvalues $\lambda_1, \lambda_2, \lambda_1 \neq \lambda_2$. Then

$$\frac{1}{\lambda_1} \int_0^T \varphi_1(t)\varphi_2(t) \, dt = \iint_0^T \varphi_1(t) r(t,\tau)\varphi_2(\tau) \, d\tau \, dt$$

$$= \frac{1}{\lambda_2} \int_0^T \varphi_1(t)\varphi_2(t) \, dt \tag{5}$$

by the symmetry of the kernel $r(t,\tau)$. This cannot hold unless the eigenfunctions are orthogonal, that is,

$$\int_0^T \varphi_1(t)\varphi_2(t) \, dt = 0. \tag{6}$$

It is possible to have several eigenfunctions corresponding to one eigenvalue $\lambda$. However, as we shall later see, there can be at most a finite number of linearly independent eigenfunctions corresponding to one eigenvalue (see [1]). Further it will be convenient to assume that the eigenfunctions have been orthonormalized by, for example, the Gramm-Schmidt procedure (see [1]).

The following theorem on the integral equation (2) will be used in the representation of the process $X_t$. *The eigenvalues $\lambda_i$ of equation (2) are countable in number. Each eigenvalue has at most a finite number of linearly independent eigenfunctions $\varphi_i(t)$ (assumed orthonormal). Furthermore, the kernel $r(t,\tau)$ has the following representation in terms of the eigenfunctions $\varphi_i(t)$ and the corresponding eigenvalues $\lambda_i$*

$$r(t,\tau) = \sum_{i=1}^{\infty} \frac{\varphi_i(t)\varphi_i(\tau)}{\lambda_i}. \tag{7}$$

This result will be derived later on. Actually the convergence in (7) is uniform.

The expansion (7) will now be used to obtain a representation of the process $X_t$, $0 \le t \le T$, in terms of the eigenfunctions and eigenvalues of equation (2). Now

$$\int_0^T E|X_t(w)|\, dt \le \int_0^T \{r(t,t)\}^{1/2}\, dt < \infty. \tag{8}$$

Fubini's theorem (see section a of Chapter IV) then implies that

$$\int_0^T |X_t(w)|\, dt \tag{9}$$

is finite for almost every realization of the process. Thus integrals of the form

$$\int_0^T X_t(w)\varphi_i(t)\, dt \tag{10}$$

are all well defined. Let

$$Z_i = \sqrt{\lambda_i} \int_0^T X_t \varphi_i(t)\, dt. \tag{11}$$

The random variables $Z_i$ are orthonormal since

$$EZ_i Z_j = \iint_0^T \varphi_i(t)\varphi_j(\tau) r(t,\tau)\, dt\, d\tau \sqrt{\lambda_i \lambda_j}$$

$$= \int_0^T \varphi_i(t)\varphi_j(t)\, dt (\lambda_j/\lambda_i)^{1/2} = \delta_{i,j} \tag{12}$$

where $E(X_t X_\tau) = r(t,\tau)$. Further, using this property,

$$E|X_t - \sum_{i=1}^{N} Z_i \varphi_i(t)/\sqrt{\lambda_i}|^2$$

$$= r(t,t) - 2 \sum_{i=1}^{N} \varphi_i(t) E(X_t Z_i)/\sqrt{\lambda_i} + \sum_{i=1}^{N} \varphi_i^2(t)/\lambda_i \qquad (13)$$

$$= r(t,t) - \sum_{i=1}^{N} \varphi_i^2(t)/\lambda_i$$

since

$$E(X_t Z_i) = \sqrt{\lambda_i} \int_0^T r(t,\tau)\varphi_i(\tau) \, d\tau = \varphi_i(t)/\sqrt{\lambda_i}. \qquad (14)$$

It follows from (7) that

$$X_t = \sum_{i=1}^{\infty} Z_i \varphi_i(t)/\sqrt{\lambda_i} \qquad (15)$$

*in the sense of mean square convergence* since (13) approaches zero as $N \rightarrow \infty$. Notice that *the $Z_i$ are independent normal variables with mean zero and variance one if $X_t$ is a Gaussian process.*

Let us now derive the results on the integral equation (2) that were assumed in the previous discussion. Let the $\varphi_i(t)$ $i = 1, \ldots, n$ be any finite number of orthonormal eigenfunctions of (2) and the $\lambda_i$ the corresponding eigenvalues. Set

$$s_n(t,\tau) = r(t,\tau) - \sum_{i=1}^{n} \varphi_i(t)\varphi_i(\tau)/\lambda_i = r(t,\tau) - q_n(t,\tau). \qquad (16)$$

Any square integrable function $g$ can be expanded in the form

$$g(t) = \sum_{i=1}^{n} a_i \varphi_i(t) + h(t) \qquad (17)$$

where $h$ is orthogonal to $\varphi_1(t), \ldots, \varphi_n(t)$. It follows that

$$\int\int_0^T g(t) s_n(t,\tau) g(\tau) \, dt \, d\tau = \int\int_0^T h(t) r(t,\tau) h(\tau) \, dt \, d\tau \geq 0. \qquad (18)$$

Let $g(t)$ equal $1/\varepsilon$ ($\varepsilon > 0$) in an interval of length $\varepsilon$ about $t_0$ and zero elsewhere. The function $s_n(t,\tau)$ is continuous in $t$ and $\tau$ jointly since $r(t,\tau)$ and the eigenfunctions $\varphi_i$ are continuous. On letting $\varepsilon \rightarrow 0$ in (18) we therefore obtain

$$s_n(t_0,t_0) = r(t_0,t_0) - \sum_{i=1}^{n} \varphi_i^2(t_0)/\lambda_i \geq 0. \qquad (19)$$

Integrate the inequality (19) from zero to $T$ to obtain

$$\int_0^T r(t,t) \, dt \geq \sum_{i=1}^{n} 1/\lambda_i. \tag{20}$$

First of all it is now clear that *there can be at most a finite number of eigenfunctions corresponding to one eigenvalue* $\lambda$. For if there were an infinite number, all the $\lambda_i$ in (20) could be taken equal to this eigenvalue $\lambda$ and the right-hand side of the inequality would be unbounded leading to a contradiction. Further *there can be at most a countable number of eigenvalues.* If there were an uncountable number of eigenvalues, an infinite number of them would have to lie in one of the sets $(1, \infty)$, $(\tfrac{1}{2},1)$, $\cdots$, $(\tfrac{1}{2}^{n+1},\tfrac{1}{2}^n)$, $\ldots$ If the $\lambda_i$ in (20) were set equal to eigenvalues lying in this particular set, the right-hand side of the inequality would be unbounded again leading to a contradiction. It is therefore seen that

$$\sum_1^\infty \varphi_i^2(t)/\lambda_i \leq r(t,t) < \infty. \tag{21}$$

From this it follows that

$$\left| \sum_{i=k}^\infty \varphi_i(t)\varphi_i(\tau)/\lambda_i \right| \leq \left[ \sum_{i=k}^\infty \varphi_i^2(t)/\lambda_i \sum_{i=k}^\infty \varphi_i^2(\tau)/\lambda_i \right]^{1/2} \tag{22}$$

so that

$$\sum_1^\infty \varphi_i(t)\varphi_i(\tau)/\lambda_i = q(t,\tau) \tag{23}$$

is convergent. Actually (22) indicates that the convergence is uniform in $\tau$ for fixed $t$ so that (23) is continuous in $\tau$ for fixed $t$. Of course, the preceding statement is valid with $\tau$ and $t$ interchanged.

We shall now introduce notation and definitions that will be helpful in the following discussion. Given any two functions $f$, $g$ that are square integrable on $0 \leq t \leq T$, let $(f,g)$ denote the "*inner product*" of the two functions where

$$(f,g) = \int_0^T f(t)g(t) \, dt. \tag{24}$$

It is immediately clear that

$$(f,g) = (g,f) \tag{25}$$

and

$$(\alpha f + \beta h, g) = \alpha(f,g) + \beta(h,g) \tag{26}$$

for any square integrable functions $f$, $h$, $g$ and real numbers $\alpha$, $\beta$. Let $\|f\|$ denote the "length" of a square integrable function where

$$\|f\| = [(f,f)]^{1/2} = \left\{ \int_0^T |f(t)|^2 \, dt \right\}^{1/2}. \tag{27}$$

The Schwarz inequality gives us a bound on the inner product $(f,g)$ in terms of the lengths $\|f\|$, $\|g\|$ of $f$, $g$

$$|(f,g)| \leq \|f\| \cdot \|g\|. \tag{28}$$

The Minkowski inequality can then be rewritten as

$$\|f + g\| \leq \|f\| + \|g\| \tag{29}$$

in this notation where $f$, $g$ are understood to be square integrable. The terms "inner product" and "length" used imply that the square integrable functions can be regarded as vectors. This is in fact the case. They are, however, vectors in an infinite dimensional space. The same notation, at the risk of hopefully small confusion, will be used for a function of two variables at times. Thus, if $r(t,\tau)$ and $s(t,\tau)$ are square integrable functions on $0 \leq t,\tau \leq T$, $(r,s)$ and $\|r\|$ will be understood to be

$$\int\!\!\!\int_0^T r(t,\tau)s(t,\tau) \, dt \, d\tau \tag{30}$$

and

$$\left\{ \int\!\!\!\int_0^T |r(t,\tau)|^2 \, dt \, d\tau \right\}^{1/2} \tag{31}$$

respectively.

*Let $r$ be square integrable on $0 \leq t, \tau \leq T$.* Then the integrable operator $R$ generated by $r$

$$(Rg)(t) = \int_0^T r(t,\tau)g(\tau) \, d\tau = h(t) \tag{32}$$

takes square integrable functions $g$ into square integrable functions $h$ on $0 \leq t \leq T$. For by the Schwarz inequality

$$\|h\| = \|Rg\| \leq \|r\| \cdot \|g\|. \tag{33}$$

In fact, it is a bounded operator in the sense that

$$\|R\| = \sup_{\|g\|>0} \frac{\|Rg\|}{\|g\|} = \sup_{\|g\|=1} \frac{\|Rg\|}{\|g\|} \leq \|r\| < \infty. \tag{34}$$

The constant $\|R\|$ is called the norm of the transformation $R$. We shall show that given any sequence of functions $g_n$ bounded in mean square,

$$\|g_n\| \leq K < \infty, \tag{35}$$

the sequence of functions $Rg_n = h_n$ contains a subsequence $h_{n_k}$ that converges in mean square

$$\|h_{n_k} - h_{n_j}\| \to 0 \tag{36}$$

as $k, j \to \infty$. Of course, the subsequence converges to a square integrable function $h(t)$ by the Riesz-Fisher theorem. This property of an operator is sometimes referred to as complete continuity. Notice that *the sum of two integral operators of type* (32) *that are completely continuous also has this property.* A simple operator of type (32) with $r(t,\tau)$ of the form

$$r_1(t,\tau) = \alpha(t)\beta(\tau), \tag{37}$$

where $\alpha$, $\beta$ are square integrable on $[0,T]$, is completely continuous. For

$$(R_1 g)(t) = (\beta,g)\alpha(t). \tag{38}$$

If $g_n$ is a sequence of functions bounded in mean square, the sequence of numbers $(\beta,g_n)$ is bounded and therefore has a finite limit point $c$. Let $g_{n_k}$ be a subsequence of $g_n$ such that $(\beta,g_{n_k}) \to c$ as $k \to \infty$. If we take $h(t) = c\alpha(t)$, it is clear that

$$\|R_1 g_{n_k} - h\| = \|\alpha\| \cdot |(\beta,g_{n_k}) - c| \to 0 \tag{39}$$

as $k \to \infty$. Thus, the very simple kernel $r(t,\tau)$ of the form (37) generates a completely continuous operator. Further it is immediately clear that a kernel $r(t,\tau)$ which is the sum of a finite number of terms of the form (37), namely

$$r_N(t,\tau) = \sum_{i=1}^{N} \alpha_i(t)\beta_i(\tau), \tag{40}$$

generates a completely continuous integral operator. Consider now a general square integrable function $r(t,\tau)$ on $0 \leq t, \tau \leq T$. The Fourier series of $r(t,\tau)$ in $t, \tau$ on $0 \leq t, \tau \leq T$ converges to $r(t,\tau)$ in mean square (see the remarks on mean square approximation of functions in section a of Chapter IV). Let $r_N(t,\tau)$ be a truncation of the Fourier series with a finite number of terms and such that

$$\|r - r_N\| \leq 1/N. \tag{41}$$

Since $r_N$ has only a finite number of terms, it is of the form (40). Let $R_N$ be the integral operator generated by $r_N$. We have already shown that the operators $R_N$ are all completely continuous. Our object is now to show that the operator $R$ they approximate is also completely continuous. Let $g_n$ be a sequence of functions bounded in mean square. Let $g_n^{(1)}$ be a subsequence of $g_n$ such that $R_1 g_n^{(1)}$ converges in mean square. Take $g_n^{(2)}$ a subsequence of $g_n^{(1)}$ such that $R_2 g_n^{(2)}$ converges in mean square. Generally, let $g_n^{(k)}$ be a subsequence of $g_n^{(k-1)}$ such that $R_k g_n^{(k)}$ converges in mean square. We derive a sequence $g^{(k)}$ by taking the $k$-th element of the $k$-th sequence. Notice that $R_N g_k^{(k)}$ converges in mean square for every $N$. This implies that $R g_k^{(k)}$ converges in mean square for

$$\|R g_k^{(k)} - R g_j^{(j)}\|$$
$$\leq \|(R - R_N) g_k^{(k)}\| + \|R_N g_k^{(k)} - R_N g_j^{(j)}\| + \|(R - R_N) g_j^{(j)}\|$$
$$\leq \|r - r_N\| (\|g_k^{(k)}\| + \|g_j^{(j)}\|) + \|r_N\| \cdot \|g_k^{(k)} - g_j^{(j)}\| \qquad (42)$$

and (42) can be made as small as is desired by taking $N$, $k$, $j$ sufficiently large. Therefore $R g_k^{(k)}$ converges in mean square to some function, say $h(t)$.

Consider the quadratic form $(Rf,f)$. Let

$$N_R = \sup_{\|f\|=1} |(Rf,f)|. \qquad (43)$$

Since

$$|(Rf,f)| \leq \|Rf\| \cdot \|f\| \qquad (44)$$

it follows that

$$N_R \leq \|R\|. \qquad (45)$$

*Assume* now *that $r(t,\tau)$ is symmetric*, that is, $r(t,\tau) = r(\tau,t)$. We will show that in this case

$$N_R = \|R\|. \qquad (46)$$

Let $\lambda > 0$. Then

$$\|Rf\|^2 = (Rf,Rf)$$
$$= \frac{1}{4}\left[\left(R\left\{\lambda f + \frac{1}{\lambda} Rf\right\}, \lambda f + \frac{1}{\lambda} Rf\right) - \left(R\left\{\lambda f - \frac{1}{\lambda} Rf\right\}, \lambda f - \frac{1}{\lambda} Rf\right)\right]$$
$$\leq \frac{1}{4} N_R\left[\left\|\lambda f + \frac{1}{\lambda} Rf\right\|^2 + \left\|\lambda f - \frac{1}{\lambda} Rf\right\|^2\right]$$
$$= \frac{1}{2} N_R\left[\lambda^2 \|f\|^2 + \frac{1}{\lambda^2} \|Rf\|^2\right]. \qquad (47)$$

The minimum of the last expression is assumed by setting $\lambda^2 = \|Rf\|/\|f\|$ if $\|Rf\| \neq 0$. The inequalities

$$\|Rf\|^2 \leq N_R \|Rf\| \cdot \|f\|, \quad \|Rf\| \leq N_R \|f\| \qquad (48)$$

then follow. Notice that these inequalities hold even if $\|Rf\| = 0$. But they imply that $\|R\| \leq N_R$ so that the equality (46) is established.

*Assume that* $\|R\| > 0$. We shall show that $R$ then has at least one nontrivial eigenfunction $\varphi$ with eigenvalue $\lambda \neq 0$. Let $f_n$ be a sequence of square integrable functions with

$$\|f_n\| = 1, \ |(Rf_n, f_n)| \to \|R\| = N_R \text{ as } n \to \infty. \tag{49}$$

We can already assume that they have been chosen so that

$$(Rf_n, f_n) \to \lambda = \|R\| \tag{50}$$

(if $R$ is positive definite). Then

$$0 \leq \|Rf_n - \lambda f_n\|^2 = \|Rf_n\|^2 - 2\lambda(Rf_n, f_n) + \lambda^2\|f_n\|^2. \tag{51}$$

The right-hand side of (51) tends to zero as $n \to \infty$ since

$$\|Rf_n\| \leq \|R\| = \lambda, \ (Rf_n, f_n) \to \lambda. \tag{52}$$

Thus

$$Rf_n - \lambda f_n \to 0 \tag{53}$$

in mean square. By the complete continuity of $R$, there is a convergent subsequence $Rf_{n_k}$ of $Rf_n$ that converges in mean square. And then by (53) it follows that $f_{n_k}$ itself has a limit in mean square, say $\varphi$. Therefore

$$R\varphi = \lambda\varphi, \tag{54}$$

that is, $\varphi$ is an eigenfunction of (2) with eigenvalue $\lambda$.

Consider the kernel $s(t, \tau) = r(t, \tau) - q(t, \tau)$. It is symmetric and square integrable. Let the corresponding integral operator be $S$. Notice that $S\varphi_i = 0$ for all the eigenfunctions $\varphi_i$ of $R$. Consider any square integrable function $f$. It has an expansion

$$f = \sum_i (f, \varphi_i)\varphi_i + h \tag{55}$$

where $(h, \varphi_i) = 0$ for all $i$. Therefore

$$Sf = Sh. \tag{56}$$

Now we must have $Sh = 0$ for all such $h$ for otherwise the norm $\|S\|$ would be positive and $S$ would have a nontrivial eigenfunction among the square integrable functions $h$ orthogonal to the $\varphi_i$'s. But such an eigenfunction would also have to be an eigenfunction of $R$ leading us to a contradiction since the $\varphi_i$'s are supposed to be all the eigenfunctions of $R$. Therefore $Sf = 0$ for all square integrable $f$. We have previously shown that continuity of $r(t,\tau)$ implies continuity of $q(t,\tau)$ in $\tau$ for fixed $t$. It follows that $s(t,\tau)$ is continuous in $\tau$ for fixed $t$. But then

$$Sf = \int_0^T [r(t,\tau) - q(t,\tau)]f(\tau)\,d\tau = 0 \qquad (57)$$

for any continuous $f$. Taking $f(\tau) = r(t,\tau) - q(t,\tau)$ we see that $r(t,\tau) - q(t,\tau) = 0$. Thus, the equality (7) is established. The uniform convergence of the series follows from the fact that the truncated series

$$\sum_{i=1}^{n} \frac{\varphi_i(t)\varphi_i(\tau)}{\lambda_i} \qquad (58)$$

which is continuous in $t$ and $\tau$ converges to a continuous function $r(t,\tau)$ as follows. It is clear that

$$\left| \sum_{i=1}^{n} \frac{\varphi_i(t)\varphi_i(\tau)}{\lambda_i} - r(t,\tau) \right| = \left| \sum_{i=n+1}^{\infty} \frac{\varphi_i(t)\varphi_i(\tau)}{\lambda_i} \right|$$

$$\leq \left\{ \sum_{i=n+1}^{\infty} \varphi_i^2(t)/\lambda_i \sum_{i=n+1}^{\infty} \varphi_i^2(\tau)/\lambda_i \right\}^{1/2}.$$

Now $\sum_1^n \varphi_i^2(t)/\lambda_i$ is continuous and converges to $r(t,t)$ monotonically. By a result called Dini's theorem (see [1]) $\sum_1^n \varphi_i^2(t)/\lambda_i$ converges to $r(t,t)$ uniformly. But this implies that $\sum_{n+1}^{\infty} \varphi_i^2(t)/\lambda_i$ converges to zero uniformly. Thus, the series (58) converges to $r(t,\tau)$ uniformly as $n \to \infty$.

An interesting illustration of the basic theorem (15) of this section is provided by the Wiener process $X_t$ on $0 \leq t \leq 1$. As we have already noted in section d of Chapter IV, the Wiener process is a normal process with covariance function

$$r(t,\tau) = \min(t,\tau), \ 0 \leq t, \tau \leq 1. \qquad (59)$$

The integral equation (2) is

$$\varphi(t) = \lambda \int_0^1 \min(t,\tau)\varphi(\tau)\,d\tau$$

$$= \lambda \left\{ \int_0^t \tau\varphi(\tau)\,d\tau + \int_t^1 t\varphi(\tau)\,d\tau \right\}. \qquad (60)$$

Equation (60) implies that

$$\varphi(0) = 0 \tag{61}$$

for every eigenfunction $\varphi$. On differentiating the equation with respect to $t$ we obtain the relation

$$\varphi'(t) = \lambda \int_t^1 \varphi(\tau)\, d\tau. \tag{62}$$

This in turn implies that

$$\varphi'(1) = 0. \tag{63}$$

The differential equation for $\varphi$

$$\varphi''(t) + \lambda\varphi(t) = 0 \tag{64}$$

is obtained by differentiating (62). The solutions of (62) are of the form

$$\varphi(t) = A \sin \sqrt{\lambda}\, t + B \cos \sqrt{\lambda}\, t. \tag{65}$$

The restraint (61) implies that $B = 0$ and the second restraint (63) implies that $\lambda = (k + \frac{1}{2})^2\pi^2$. The orthonormalized eigenfunctions of (60) are therefore

$$\varphi_k(t) = \sqrt{2} \sin (k + \tfrac{1}{2})\pi t \tag{66}$$

with corresponding eigenvalues $\lambda_k = (k + \frac{1}{2})^2\pi^2$, $k = 1, 2, \ldots$. The Wiener process $X_t$ on $0 \le t \le 1$ therefore has the representation

$$X_t = \sqrt{2} \sum_{k=1}^{\infty} Z_k \sin (k + \tfrac{1}{2})\pi t / (k + \tfrac{1}{2})\pi. \tag{67}$$

## d. A Uniform Mixing Condition and Narrow Band-Pass Filtering

The concept of mixing was introduced in section b of Chapter V. This property is appreciably stronger than that of ergodicity. However, a central limit theorem need not hold for a dependent process satisfying even this requirement. We shall discuss here the property of uniform mixing which was introduced in [66], [70] and is strong enough to imply a central limit theorem.

Let $X_t(w) = X_t$, $-\infty < t < \infty$, be a random process (not necessarily stationary) that is measurable in $t$ and $w$ jointly. Let $\mathfrak{B}_t$ be the Borel field of events generated by the random variables $X_u$, $u \le t$, and $\mathfrak{F}_\tau$ the Borel field of events generated by the random variables $X_u$, $u \ge \tau$. Thus $\mathfrak{B}_t$ and $\mathfrak{F}_\tau$ represent the information given by knowledge of the process before $t$ and after $\tau$ respectively. *The process $X_t$ satisfies a uniform mixing condition if there is some positive function $g(s)$ defined for $0 \leq s < \infty$ with $g(s) \to 0$ as $s \to \infty$ such that for any pair of events $B\epsilon\mathfrak{B}_t$, $F\epsilon\mathfrak{F}_\tau$, $t < \tau$,*

$$|P(B \cap F) - P(B)P(F)| < g(\tau - t). \tag{1}$$

Further any process $Y_t$ derived from the $X_t$ process by operations over a finite time interval and their shifts is also uniformly mixing. Specifically, if for some fixed $L$, $0 < L < \infty$, and every $t$ $Y_t(w)$ is measurable with respect to the Borel field generated by $X_u(w)$, $t - L \leq u \leq t$, it follows that the process $Y_t(w)$ is uniformly mixing. Asymptotic normality for time averages of a discrete parameter uniformly mixing process was obtained in [66] under certain moment conditions. However, the proof can easily be modified so as to get the corresponding result for continuous parameter processes. Recently Kolmogorov and Rosanov obtained conditions under which a stationary Gaussian process is uniformly mixing [47].

*Consider a process $X_t$, $-\infty < t < \infty$, with finite fourth moments, $EX_t^4 < \infty$, and continuous in the mean of fourth order, that is,*

$$E|X_t - X_\tau|^4 \to 0 \tag{2}$$

as $t \to \tau$. It will be convenient to *take the first moment $EX_t$ as identically zero.* Further *assume* that $X_t$ is *stationary in moments of second order and fourth order* so that

$$E[X_{t_1}X_{t_2}] = r(t_2 - t_1)$$
$$E[X_{t_1}X_{t_2}X_{t_3}X_{t_4}] = P(t_2 - t_1, t_3 - t_1, t_4 - t_1). \tag{3}$$

Introduce the fourth-order cumulant function

$$Q(t_2 - t_1, t_3 - t_1, t_4 - t_1) = P(t_2 - t_1, t_3 - t_1, t_4 - t_1)$$
$$- P_G(t_2 - t_1, t_3 - t_1, t_4 - t_1) \tag{4}$$

where $P_G$ is what $P$ would be in the case of a normal process, namely,

$$P_G(t_2 - t_1, t_3 - t_1, t_4 - t_1) = r(t_2 - t_1)r(t_4 - t_3)$$
$$+ r(t_3 - t_1)r(t_4 - t_2) + r(t_4 - t_1)r(t_3 - t_1) \tag{5}$$

(see [54]). Notice that if $r(t)$ is absolutely integrable, the spectral density $f(\lambda)$ of the process $X_t$ exists and is continuous. We shall *assume that $r(t)$ and $Q(t_1,t_2,t_3)$ are absolutely integrable over one- and three-dimensional space respectively.* Further *the spectral density $f(\lambda)$ will be assumed positive everywhere.*

In the communication engineering literature it is often assumed that a narrow band-pass filter applied to a stationary random input yields an output that is approximately normally distributed. Such an output is of the form

$$\int_0^T w_T(t) X_t \, dt \qquad (6)$$

where the weight function $w_T(t)$ is chosen so that the spectral mass of $X_t$ away from $\lambda$, $-\lambda$ ($\lambda$ the frequency of interest) is damped out and that around $\lambda$, $-\lambda$ is passed through. We shall consider a class of weight functions $w_T(t)$ large enough to include such filters. Let

$$W(T) = \int_0^T w_T^2(t) \cdot dt. \qquad (7)$$

Assume that the weight functions $w_T(t)$ satisfy the following conditions:

1. $W(T) \to \infty$ as $T \to \infty$                           (8)
2. *The functions* $W_T(t)$ *are slowly increasing in that*

(a) $\displaystyle\int_{A(T)} |w_T(t)|^2 \, dt = o(W(T))$ *as* $T \to \infty$     (9)

*for any sequence of subsets* $A(T)$ *of* $[0,T]$ *with the Lebesgue measure* (of $A(T)$) $m(A(T)) = o(T)$, *uniformly in* $m(A(T))/T$ *as* $T \to \infty$ *and*

(b) $w_T(t) = o(W(T)^{1/2})$ *uniformly in* $t$ *as* $T \to \infty$     (10)

3. $\displaystyle\lim_{T \to \infty} W(T)^{-1} \int_0^{T-|h|} w_T(t+|h|) w_T(t) \, dt = \rho(h)$     (11)

*exists for every* $h$ *and is continuous in* $h$. The limit function $\rho(h)$ can be seen to be a positive definite function and therefore has a representation

$$\rho(h) = \int_{-\infty}^{\infty} e^{ih\lambda} \, dM(\lambda) \qquad (12)$$

with $M(\lambda)$ a nondecreasing bounded function by Bochner's theorem. Notice that $dM(\lambda) = dM(-\lambda)$ since $\rho(h) = \rho(-h)$. When $w_T(t)$ corresponds to a narrow band-pass filter, conditions 1 and 2 of this section will usually be obviously satisfied since the functions $w_T(t)$ are typically uniformly bounded in $T$ with $W(T)$ the same order of magnitude as $T$. In the case of narrow band-pass filtering about $\mu$, $M(\lambda)$ will increase only at $\mu$ and $-\mu$. Actually the conditions on $w_T(t)$ are general enough to cover situations that often occur in regression problems as they arise in time series analysis [26].

Let $X_t$, $EX_t \equiv 0$, be a uniformly mixing process stationary up to moments of fourth order and satisfying the conditions on $r(t)$ and $Q(t_1,t_2,t_3)$ specified before. Further let the weight function $w_T(t)$ satisfy the assumptions 1, 2, 3 of this section. We shall then show that

$$W(T)^{-\frac{1}{2}} \int_0^T w_T(t) X_t \, dt \tag{13}$$

is asymptotically normally distributed with mean zero and variance

$$2\pi \int_{-\infty}^{\infty} f(\lambda) \, dM(\lambda). \tag{14}$$

Our proof will follow that given in [70].

Let

$$S(T) = \int_0^T w_T(t) X_t \, dt \tag{15}$$

and set

$$U_j(T) = \int_{j[p(T)+q(T)]}^{(j+1)p(T)+jq(T)} w_T(t) X_t \, dt \tag{16}$$

$$V_j(T) = \int_{(j+1)p(T)+jq(T)}^{(j+1)[p(T)+q(T)]} w_T(t) X_t \, dt \tag{17}$$

$j = 0, 1, \ldots, k-1$. Here $k[p(T) + q(T)] = T$ and $k = k(T)$, $p(T)$, $q(T)$ will be chosen so that $k$, $p$, $q \to \infty$ and $q(T)/p(T) \to 0$ as $T \to \infty$. The interval $[0,T]$ is being divided into an alternating succession of big blocks and small blocks, each of length $p(T)$ and $q(T)$ respectively. The $U_j$'s and $V_j$'s are the large and small block integrals respectively. We first show that the contribution of the small block integrals is negligible. Now

$$E \left| \sum_{j=1}^k V_j W(T)^{-\frac{1}{2}} \right|^2 = E \left| \int_{A(T)} w_T(t) X_t \, dt \, W(T)^{-\frac{1}{2}} \right|^2$$

$$\leq \int |r(u)| \int_{A(T)} |w_T(t+u) w_T(t)| \, dt \, du \, W(T)^{-1}$$

$$\leq \int |r(u)| \left[ \int_{A(T)} |w_T(t+u)|^2 \, dt \int_{A(T)} |w_T(t)|^2 \, dt \right]^{\frac{1}{2}} W(T)^{-1} \tag{18}$$

where

$$A(T) = \bigcup_{j=1}^k \{t \mid jp(T) + (j-1)q(T) \leq t \leq j(p(T)+q(T))\}. \tag{19}$$

But (18) must approach zero because of the absolute integrability of

$r(u)$ and condition 2 of this section. But then

$$\sum_{1}^{k} V_j W(T)^{-\frac{1}{2}} \to 0 \qquad (20)$$

in probability as $T \to \infty$.

Our object now is to show that

$$\sum_{1}^{k} U_j W(T)^{-\frac{1}{2}} \qquad (21)$$

is asymptotically normally distributed with mean zero and variance (14) as $T \to \infty$. The theorem on asymptotic normality of $S(T)W(T)^{-\frac{1}{2}}$ would then follow immediately for (20) is asymptotically negligible. Introduce for this purpose the distribution functions

$$G_{j,T}(x) = P\{U_j W(T)^{-\frac{1}{2}} \le x\} \qquad (22)$$

and the events

$$A(j,T,m_j,\delta) = \{m_j\delta < U_j W(T)^{-\frac{1}{2}} \le (m_j + 1)\delta\}. \qquad (23)$$

Now with $m_1, \ldots, m_k$ integers

$$\sum_{(m_1 + \cdots + m_k + k)\delta \le x} P(\bigcap_{j=1}^{k} A(j,T,m_j,\delta)) \le P(\sum_{j=1}^{k} U_j W(T)^{-\frac{1}{2}} \le x)$$

$$\le \sum_{(m_1 + \cdots + m_k)\delta \le x} P(\bigcap_{j=1}^{k} A(j,T,m_j,\delta)). \qquad (24)$$

Notice that

$$P(\max_{j=1,\ldots,k} |U_j W(T)^{-\frac{1}{2}}| > \tau_k) < \varepsilon \qquad (25)$$

where $\tau_k = k(C/\varepsilon)^{\frac{1}{2}}$ with $C$ a constant. This follows from the fact that

$$E|\max_{j=1,\ldots,k} |U_j W(T)^{-\frac{1}{2}}||^2 \le E\sum_{1}^{k} |U_j W(T)^{-\frac{1}{2}}||^2$$

$$\le W(T)^{-1}(\sum_{j=1}^{k} E^{\frac{1}{2}}|U_j|^2)^2 \le k^2 C \qquad (26)$$

and an application of the Chebyshev inequality.

Further

$$\left| \sum_{(m_1 + \cdots + m_k)\delta \le x} P(\bigcap_{1}^{k} A(j,T,m_j,\delta)) - \sum_{(m_1 + \cdots + m_k)\delta \le x} \prod_{1}^{k} P(A(j,T,m_j,\delta)) \right|$$

$$\le k\left(\frac{2\tau_k}{\delta}\right)^k g(q(T)) + 2\varepsilon. \qquad (27)$$

For the probability contributed by all the sets $\bigcap_{j=1}^{k} A(j,T,m_j,\delta)$ for which

$\max |U_j W(T)^{-\frac{1}{2}}| > \tau_k$ is by (25) at most $\varepsilon$. Consider all the sets $\bigcap_{j=1}^{k} A(j,T,m_j,\delta)$ for which $\max |U_j W(T)^{-\frac{1}{2}}| \le \tau_k$. By repeated application of the uniform mixing condition it is clear that

$$\left| P(\bigcap_{j=1}^{k} A(j,T_j,m,\delta)) - \prod_{j=1}^{k} P(A(j,T,m_j,\delta)) \right| \le kg(q(T)). \quad (28)$$

Since there are $(2\tau_k/\delta)^k$ sets of this form, the desired inequality (27) is obtained.

Inequality (27) will be applied later to show that the sum of the $U_j W(T)^{-\frac{1}{2}}$ has the same asymptotic distribution as the sum of independent random variables with the same marginal distributions, as long as $k(T)$, $q(T)$, $p(T)$ are appropriately chosen. But first let us see what the asymptotic distribution of the sum of such independent random variables would be. The distribution function of the sum of these $k$ independent random variables is

$$G_{1,T}* \ \cdot \ \cdot \ \cdot \ *G_{k,T}(x), \quad (29)$$

the convolution of $G_{1,T}(x), \ldots, G_{k,T}(x)$. We now show that (21) *is asymptotically normally distributed with mean zero and variance* (14). Now

$$\sum_{j=1}^{k} E|U_j W(T)^{-\frac{1}{2}}|^2 \to \int_{-\infty}^{\infty} r(u)\rho(u) \, du = 2\pi \int_{-\infty}^{\infty} f(\lambda) \, dM(\lambda) \quad (30)$$

as $T \to \infty$ by conditions 2a and 3 of this section. Further (30) is positive since $f(\lambda)$ is positive everywhere. By Lyapunov's form of the central limit theorem (see Loève [53]), if

$$\sum_{j=1}^{k} E|U_j W(T)^{-\frac{1}{2}}|^4 = o(\sum_{j=1}^{k} E|U_j W(T)^{-\frac{1}{2}}|^2)^2 \quad (31)$$

expression (21) is asymptotically normal with mean zero and variance (14). But

$$E|U_j W(T)^{-\frac{1}{2}}|^4 = W(T)^{-2} \int\!\!\int\!\!\int\!\!\int_{j[p(T)+q(T)]}^{(j+1)p(T)+jq(T)} w_T(t_1)w_T(t_2)w_T(t_3)w_T(t_4)$$
$$\cdot [r(t_1 - t_2)r(t_3 - t_4) + r(t_1 - t_3)r(t_2 - t_4) + r(t_1 - t_4)r(t_2 - t_3)$$
$$+ Q(t_2 - t_1, t_3 - t_1, t_4 - t_1)] \, dt_1 \, dt_2 \, dt_3 \, dt_4. \quad (32)$$

The sum over $j$ of the first three terms on the right-hand side of equality (32) is

$$3 \sum_{j=1}^{k} E^2|U_j W(T)^{-\frac{1}{2}}|^2. \quad (33)$$

By condition 2a, $E|U_jW(T)^{-\frac{1}{2}}|^2$ approaches zero uniformly in $j$. This coupled with (30) implies that expression (33) approaches zero as $T \to \infty$. Consider the last term on the right-hand side of (32). By condition 2 and the absolute integrability of $Q$, the sum over $j$ of the last term on the right of (32) tends to zero as $T \to \infty$. Lyapunov's condition for the central limit theorem is therefore satisfied.

Notice that

$$G_{1,T*} \cdots *G_{k,T}(x - k\delta) \leq \sum_{(m_1 + \cdots + m_k + k)\delta \leq x} \prod_{j=1}^{k} P(A(j,T,m_j,\delta))$$

$$\leq \sum_{(m_1 + \cdots + m_k)\delta \leq x} \prod_{j=1}^{k} P(A(j,T,m_j,\delta)) \leq G_{1,T*} \cdots *G_{k,T}(x + k\delta).$$

$$(34)$$

The asymptotic normality of (29) coupled with (27), (30), (31), and (34) implies the desired theorem if $\delta(T)$, $k(T)$, $p(T)$, $q(T)$ can be chosen so that

$$k(T)[p(T) + q(T)] = T$$

$$k(T), p(T), q(T) \to \infty$$

$$q(T)/p(T) \to 0 \qquad (35)$$

$$k(T)\delta(T) \to 0$$

$$k\left(\frac{2\tau_k}{\delta}\right)^k g(q(T)) \to 0.$$

The condition $k(T)\delta(T) \to 0$ is easily satisfied if we set $\delta = k^{-2}$. The difficult condition to satisfy is the last one. Now

$$k(2\tau_k/\delta)^k \leq k^{5k}D^k \qquad (36)$$

with $D = 2(C/\varepsilon)^{\frac{1}{2}}$. Given the existence of a function $g$ satisfying (1), it can always be taken so that

$$g(u) > (u + 1)^{-1}. \qquad (37)$$

If $k$ is chosen so that

$$k \leq [-\log g(q(T))]^{\frac{1}{2}} \qquad (38)$$

the last of the conditions (35) is satisfied. Keeping these remarks in mind, it is clear that if one takes $q(T) = T^{\frac{1}{2}}$ for large $T$, all the conditions (35) can be satisfied. The proof of the asymptotic normality of (6) is now complete.

## e. Problems

1. Consider the two-state (say 0 and 1) stationary Markov chain $X_n$ with transition probability matrix $\mathbf{P} = \begin{pmatrix} p & q \\ q & p \end{pmatrix}, 0 < p < q < 1,$ $q = 1 - p$, and instantaneous probability vector $(\tfrac{1}{2},\tfrac{1}{2})$. Show that the random variables

$$\xi_n = \xi(X_n,X_{n-1}) = \begin{cases} 1 & \text{if } (X_n,X_{n-1}) = (1,1) \text{ or } (0,0) \\ 0 & \text{otherwise} \end{cases}$$

are independent and identically distributed. Further show that $X_n$ is determined by $\xi_n, \xi_{n-1}, \ldots, \xi_{n-k}, X_{n-k-1}$ for every $k = 0,$ 1, 2, . . . . Is $X_n$ determined by $\xi_n, \xi_{n-1}, \ldots$ ? If not, why not?

2. Let $X_t$ be a Gaussian process with covariance function $\varphi(t,\tau) = \min (t,\tau) - t\tau, 0 \le t, \tau \le 1$. Find the representation of $X_t$ on $[0,1]$ in terms of the eigenfunctions and eigenvalues of the integral equation with kernel $\varphi(t,\tau)$.

3. Consider the previous problem with covariance function $\varphi(t,\tau) = \exp(-|t - \tau|), 0 \le t, \tau \le 1$.

4. Consider the previous problem with covariance function

$$\varphi(t,\tau) = \Sigma \alpha_j \exp(-\beta_j|t - \tau|),$$

$0 \le t, \tau \le 1$, where the $\beta_j$'s are positive and the $\alpha_j$'s are chosen so that the function $\varphi(t,\tau)$ is positive definite.

5. Let $\xi_t$ be the Wiener process. Suppose $X_t$ is a process constructed so that $X_t$ is functionally dependent on $\xi_t - \xi_\tau$ for $t - L \le \tau \le t$ where $L$ is a fixed positive number. Is $X_t$ uniformly mixing?

6. Let $\xi_t$ be a continuous parameter $(-\infty < t < \infty)$ process with the increments of $\xi_t$ $(\xi_t - \xi_\tau)$ over disjoint intervals independent. Further, assume that $\xi_t - \xi_\tau$ $(t > \tau)$ is Poisson distributed with mean $\lambda(t - \tau)$. Examine the limiting distribution of $\dfrac{1}{T} \displaystyle\int_0^T (\eta_t - \lambda)\, dt$ where $\eta_t = \xi_t - \xi_{t-1}$ as $T \to \infty$. What can you say about it in terms of the central limit theorem derived in this chapter?

# Notes

1. The zero-one law derived in section a is commonly called the Borel-Cantelli lemma.

2. The problem considered in section b can be posed in the broader context of real valued Markov processes or strictly stationary processes. Refer to references [A7] and [67] for work in this general setting. There are still many interesting open questions. Some related work is taken up in the isomorphism problem (see, for example, D. Ornstein [A11]).

3. Most of section c is devoted to the proof of a result on integral equations called Mercer's theorem. It was felt that it would be more convenient to have a proof in the text rather than to refer the reader to a book with material on integral equations (such as [63], for example). Notice that the result on the representation of processes follows almost immediately from Mercer's theorem. The representation theorem for processes is usually attributed to Karhunen and Loève.

4. The uniform mixing condition of section d has been of increasing interest in recent years (see [A9] and [A13]).

# REFERENCES

[1] T. M. Apostol. *Mathematical Analysis.* 1957.

[2] G. K. Batchelor. *The Theory of Homogeneous Turbulence.* 1953, Cambridge.

[3] S. N. Bernstein. Démonstration du théorème de Weierstrass fondée sur le calcul de probabilité. *Proc. Kharkhov Math. Soc.* XIII, 1912.

[4] S. N. Bernstein. Equations differentielles stochastiques. *Actualités Sci. Ind.* 738 (1938), 5–31.

[5] G. Birkhoff and R. S. Varga. Reactor criticality and non-negative matrices. *J. Soc. Indust. Appl. Math.* 6 (1958), 354–77.

[6] R. B. Blackman and J. W. Tukey. *The Measurement of Power Spectra from the Point of View of Communications Engineering.* 1958, New York.

[7] C. J. Burke and M. Rosenblatt. A Markovian function of a Markov chain. *Ann. Math. Statist.* 29 (1958), 1112–22.

[8] K. L. Chung. *Markov Chains with Stationary Transition Probabilities.* 1960, Berlin.

[9] R. Courant. *Differential and Integral Calculus.* vol. 1, 1937, New York.

[10] W. Doeblin. Sur les propriétés asymptotiques de mouvement régis par certain types de chaines simples. *Bull. Math. Soc. Roum. Sci.* 39 (1937), no. 1, 57–115; no. 2, 3–61.

[11] J. L. Doob. Topics in the theory of Markov chains. *Trans. Amer. Math. Soc.* 52 (1942), 37–64

[12] J. L. Doob. *Stochastic Processes.* 1953, New York.

[13] J. L. Doob. Discrete potential theory and boundaries. *J. Math. Mech.* 8 (1959), 433–58.

[14] A. Feinstein. *Foundations of Information Theory.* 1958, New York.

[15] W. Feller. Zur Theorie der stochastischen Prozesse (Existenz und Eindeutigkeitssätze). *Math. Ann.* 113 (1936), 113–60.

[16] W. Feller. On the integro-differential equations of purely discontinuous Markov processes. *Trans. Amer. Math. Soc.* 48 (1940), 488–515.

[17] W. Feller. *An Introduction to Probability Theory and Its Applications.* Vol. 1, 1950, New York.

[18] P. Frank and R. von Mises. *Die Differential und Integralgleichungen der Mechanik und Physik.* Vol. 2, 1943, New York.

[19] M. Frechet. *Recherches Théoriques modernes sur le calcul des probabilités* II. Méthode des fonctions arbitraires. Theorie des événements en chaine dans le cas d'un nombre fini d'états possibles. 1938, Paris.

[20] W. Freiberger and U. Grenander. Approximate distributions of noise power measurements. *Quart. Appl. Math.* 17 (1959), 271–84.

[21] F. R. Gantmacher. *The Theory of Matrices.* Vols. 1, 2, 1959, New York.

[22] H. Goldstein. *Classical Mechanics.* 1950, Cambridge.

[23] B. V. Gnedenko and A. Kolmogorov. *Limit Distributions for Sums of Independent Random Variables.* 1954, Cambridge, Mass.

[24] N. R. Goodman. *On the Joint Estimation of the Spectra, Cospectrum and Quadrature Spectrum of a Two-dimensional Stationary Gaussian Process.* Scientific Paper No. 10, Engineering Statistics Laboratory, New York University (1957), p. 168.

[25] U. Grenander. On the estimation of regression coefficients in the case of an autocorrelated disturbance. *Ann. Math. Statist.* 25 (1954), 252–72.

[26] U. Grenander and M. Rosenblatt. *Statistical Analysis of Stationary Times Series.* 1957, New York.

[27] U. Grenander and G. Szegö. *Toeplitz Forms and Their Applications.* 1958, Berkeley.

[28] U. Grenander, H. O. Pollak, and D. Slepian. The distribution of quadratic forms in normal variates: a small sample theory with applications to spectral analysis. *J. Soc. Indust. Appl. Math.* 7 (1959), 374–401.

[29] P. R. Halmos. *Measure Theory.* 1950, New York.

[30] P. R. Halmos. *Lectures on ergodic theory.* Math. Soc. Japan, 1956.

[31] T. E. Harris. Branching processes. *Ann. Math. Statist.* 19 (1948), 474–94.

[32] T. E. Harris. *Some Mathematical Models for Branching Processes.* Proc. 2nd Berkeley Symp. Math. Stat. and Prob. (1951), 305–28.

[33] H. Helson and D. Lowdenslager. Prediction theory and Fourier series in several variables. *Acta Math.* 99 (1958), 165–202.

[34] E. Hille and R. S. Phillips. *Functional Analysis and Semigroups.* 1957, Providence, R.I.

[35] E. Hopf. Ergodentheorie. *Erg. Math.* 5, no. 2 (1937).

[36] K. Ito. On stochastic differential equations. *Mem. Am. Math. Soc.* 4 (1951), p. 51.

[37] M. Kac. *Probability and Related Topics in Physical Sciences.* 1959, New York.

[38] K. Karhunen. Über lineare Methoden in der Wahrscheinlichkeitsrechnung. *Ann. Acad. Sci. Fennicae,* Ser. A, I. Math. Phys. 37 (1947), p. 79.

[39] S. Karlin and J. L. McGregor. The differential equations of birth-and-death processes and the Stieltjes moment problem. *Trans. Amer. Math. Soc.* 85 (1957), 489–546.

[40] A. I. Khinchin. *Asymptotische Gesetze der Wahrscheinlichkeitsrechnung.* 1948, New York.

[41] A. I. Khinchin. *Mathematical Foundations of Statistical Mechanics.* 1949, New York.

[42] A. I. Khinchin. *Mathematical Foundations of Information Theory.* 1957, New York.

[43] A. Kolmogorov. Über die analytischen Methoden in der Wahrscheinlichkeitsrechnung. *Math. Ann.* 104 (1931), 415–58.

[44] A. Kolmogorov. *Foundations of Probability Theory.* 1956, New York.

[45] A. Kolmogorov. Anfangsgründe der Markoffschen Ketten mit unendlich vielen möglichen Zuständen. *Rec. Math. Moscow* (Mat. Sb.) 1 (43)(1936), 607–10.

[46] A. Kolmogorov. Stationary sequences in Hilbert space. *Bull. Math. Univ. of Moscow,* 2(1941).

[47] A. Kolmogorov and Iu. A. Rosanov. On a strong mixing condition for stationary Gaussian processes. *Teor. Veroyatnost. i. Primenen,* 5 (1960).

[48] T. Koopmans (ed.). *Statistical Inference in Dynamic Economic Models.* Cowles Commission for Research in Economics, Monograph No. 10, 1950, New York.

[49] E. E. Levi. Sull'equazione del calore. *Ann. di. mat. pura appl.* (3)14(1908).

[50] P. Lévy. *Processus stochastiques et mouvement Brownien.* Paris, 1948.

[51] P. Lévy. *Processus doubles de Markoff. Le calcul des probabilités et ses applications.* 1949, Paris, 53–9.

[52] P. Lévy. Examples de processus pseudo-Markoviens. *C. R. Acad. Sci.* 228 (1949), Paris, 2004–6.

[53] M. Loève. *Probability Theory.* 1955, New York.

[54] T. A. Magness. Spectral response of a quadratic device to non-Gaussian noise. *J. Appl. Phys.* 25 (1954), 1357–65.

[55] G. Maruyama. The harmonic analysis of stationary stochastic processes. *Mem. Fac. Sci. Kyushu Univ.* A4 (1949), 45–106.

[56] P. Masani and N. Wiener. Prediction theory of multivariate stochastic processes. *Acta Math.* 98 (1957), 111–150, 99 (1958), 93–137.

**References**

[57] R. von Mises. *Wahrscheinlichkeitsrechnung*. 1931, Leipzig and Wien.

[58] L. S. Ornstein and G. E. Uhlenbeck. On the theory of the Brownian motion. *Phys. Rev.* 36 (1930), 823–41.

[59] E. Parzen. On consistent estimates of the spectrum of a stationary time series. *Ann. Math. Statist.* 28 (1957), 329–48.

[60] W. J. Pierson, Jr. *Wind generated gravity waves. Advances in Geophysics.* Vol. 2, Academic Press, 1955, New York.

[61] B. Rankin. *The concept of enchainment—a relation between stochastic processes.* Unpublished (1955).

[62] F. Riesz. Sur la Théorie ergodique. *Comm. Math. Helv.* 17 (1945), 221–39.

[63] F. Riesz and B. Sz-Nagy. *Leçons d'analyse fonctionelle.* 1952, Budapest.

[64] D. Rosenblatt. On some aspects of complex behavioral systems. *Information and Decision Processes.* 1960, 62–86, New York.

[65] D. Rosenblatt. On aggregation and consolidation in linear systems. To be published in *Naval Res. Logistics Quarterly.*

[66] M. Rosenblatt. A central limit theorem and a strong mixing condition. *Proc. Nat. Acad. Sci. U.S.A.* 42 (1956), 43–7.

[67] M. Rosenblatt. Stationary processes as shifts of functions of independent random variables. *J. Math. Mech.* 8 (1959), 665–82.

[68] M. Rosenblatt. Functions of a Markov process that are Markovian. *J. Math. Mech.* 8 (1959), 585–96.

[69] M. Rosenblatt. Stationary Markov chains and independent random variables. *J. Math. Mech.* 9 (1960), 945–50.

[70] M. Rosenblatt. Some comments on narrow band-pass filters. *Quart. Appl. Math.* 18 (1961), 387–93.

[71] M. Rosenblatt. Statistical analysis of stochastic processes with stationary residuals. *H. Cramér volume*, 1960, New York, 246–75.

[72] P. Scheinok. The error on using the asymptotic variance and bias of spectrograph estimates for finite observations. Unpublished thesis, Indiana University. 1960.

[73] G. E. Shannon. The mathematical theory of communication. *Bell System Tech. J.* 27 (1948), 379–423, 623–56.

[74] A. J. W. Sommerfeld. *Partial Differential Equations in Physics.* 1949, New York.

[75] G. Szegö. Beiträge zuer Theorie der Toeplitzschen Formen I, II. *Math. Z.* 6 (1920), 167–202, 9 (1921), 167–90.

[76] H. Wielandt. Unzerlegbare nicht-negative Matrizen. *Math. Z.* 52 (1950), 642–8.

[77] N. Wiener. *Extrapolation, Interpolation and Smoothing of Stationary Time Series.* 1949, New York.

# ADDITIONAL REFERENCES

[A1] T. W. Anderson. *The Statistical Analysis of Time Series.* 1971, New York.

[A2] B. M. Brown and G. K. Eagleson. Martingale convergence to infinitely divisible laws with finite variances. *Trans. Amer. Math. Soc.*, 162 (1971), 449–453.

[A3] R. L. Dobrushin. Description of a random field by means of conditional probabilities and the conditions governing its regularity. *Theory of Prob. and Its Appl.*, 13 (1968), 197–224.

[A4] E. B. Dynkin and A. A. Yushkevich. *Markov Processes. Theorems and Problems.* New York, 1969.

[A5] S. R. Foguel. *The Ergodic Theory of Markov Processes.* 1969, New York.

[A6] M. I. Gordin. The central limit theorem for stationary processes. *Soviet Math. Doklady*, 10 (1969), 1174–1176.

[A7] E. J. Hannan. *Multiple Time Series.* 1970, New York.

[A8] D. L. Hanson. On the representation problem for stationary stochastic processes with trivial tail field. *J. Math. Mech.*, 12 (1963), 293–301.

[A9] I. A. Ibragimov and Iu. V. Linnik. *Independent and Stationary Sequences of Random Variables*, 1971, Groningen.

[A10] P. A. Meyer. *Martingales and Stochastic Integrals I.* Lecture Notes in Mathematics, 284. 1972, Berlin.

[A11] D. S. Ornstein. Bernoulli shifts with the same entropy arc isomorphic. *Advances in Math.*, 4 (1970), 337–352.

[A12] L. D. Pitt. A Markov property for Gaussian processes with a multidimensional parameter. *Arch. Rational Mech. Anal.*, 43 (1971), 367–391.

[A13] Iu. Rosanov. *Stationary Random Processes.* 1967, San Francisco.

[A14] M. Rosenblatt. *Markov Processes. Structure and Asymptotic Behavior.* 1971, Berlin.

[A15] C. Stein. A bound for the error in the normal approximation to the distribution of a sum of dependent random variables. *Proceedings of the 6th Berkeley Symposium on Mathematical Statistics and Probability.* Vol. II, 583–602. 1972, Berkeley.

# INDEX

227